Diverse Applications of Principal Component Analysis

Diverse Applications of Principal Component Analysis

Edited by **Rebecca Cross**

CLANRYE INTERNATIONAL

New Jersey

Published by Clanrye International,
55 Van Reypen Street,
Jersey City, NJ 07306, USA
www.clanryeinternational.com

Diverse Applications of Principal Component Analysis
Edited by Rebecca Cross

International Standard Book Number: 978-1-63240-150-2 (Hardback)

Printed in the United States of America.

Contents

Preface

This book emphasizes on the diverse applications of principal component analysis. The aim of this book is to enhance knowledge of scientists, engineers and researchers regarding the advantages of this technique in data analysis. It includes the uses of PCA in distinct fields like agriculture, architecture, taxonomy, pharmacy, ecology, biology, health and finance.

All of the data presented henceforth, was collaborated in the wake of recent advancements in the field. The aim of this book is to present the diversified developments from across the globe in a comprehensible manner. The opinions expressed in each chapter belong solely to the contributing authors. Their interpretations of the topics are the integral part of this book, which I have carefully compiled for a better understanding of the readers.

At the end, I would like to thank all those who dedicated their time and efforts for the successful completion of this book. I also wish to convey my gratitude towards my friends and family who supported me at every step.

Editor

Principal Component Analysis in the Era of «Omics» Data

Louis Noel Gastinel
University of Limoges, University Hospital of Limoges, Limoges
France

1. Introduction

1.1 Definitions of major «omics» in molecular biology and their goals

The «omics» era, also called classically the post-genomic era, is described as the period of time which extends the first publication of the human genome sequence draft in 2001 (International Human Genome Sequencing Consortium, 2001; Venter et al., 2001). Ten years after that milestone, extensive use of high-throughput analytical technologies, high performance computing power and large advances in bioinformatics have been applied to solve fundamental molecular biology questions as well as to find clues concerning human diseases (cancers) and aging. Principal «omics», such as Gen-*omics*, Transcript-*omics*, Prote-*omics* and Metabol-*omics*, are biology disciplines whose main and extremely ambitious objective is to describe as extensively as possible the complete class-specific molecular components of the cell. In the «omics» sciences, the catalog of major cell molecular components, respectively, genes, messenger RNAs and small interfering and regulatory RNAs, proteins, and metabolites of living organisms, is recorded qualitatively as well as quantitatively in response to environmental changes or pathological situations. Various research communities, organized in institutions both at the academic and private levels and working in the «omics» fields, have spent large amounts of effort and money to reach. standardization in the different experimental and data processing steps. Some of these «omics» specific steps basically include the following: the optimal experimental workflow design, the technology-dependent data acquisition and storage, the pre-processing methods and the post-processing strategies in order to extract some level of relevant biological knowledge from usually large data sets. Just like Perl (Practical Extraction and Report Language) has been recognized to have saved the Human Genome project initiative (Stein, 1996), by using accurate rules to parse genomic sequence data, other web-driven. programming languages and file formats such as XML have also facilitated «omics» data dissemination among scientists and helped rationalize and integrate molecular biology data.

Data resulting from different «omics» have several characteristics in common, which are summarized in Figure 1: (a) the number of measured variables n (SNP, gene expression, proteins, peptides, metabolites) is quite large in size (from 100 to 10000), (b) the number of samples or experiments p where these variables are measured associated with factors such as the pathological status, environmental conditions, drug exposure or kinetic points (temporal experiments) is rather large (10 to 1000) and (c) the measured variables are organized in a matrix of n x p dimensions. The cell contents of such a matrix usually record a metric (or numerical code) related to the abundance of the measured variables. The observed data are acquired keeping the lowest amount of possible technical and analytical variability. Exploring these «omics» data requires fast computers and state-of-the-art data visualization and statistical multivariate tools to extract relevant knowledge, and among these tools PCA is a tool of choice in order to perform initial exploratory data analysis (EDA).

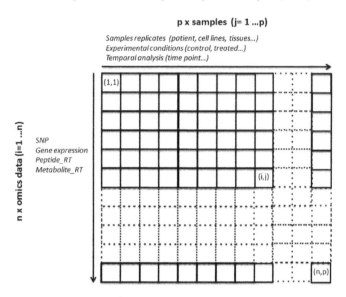

Fig. 1. General organization of raw «omics» data represented in a n x p matrix.

Rows contain the measured quantitative variables (n) and columns contain the samples or experimental conditions tested (p) from which variables n are measured and for which grouping information or factors is generally present. Each cell (i,j) of this matrix contains a measured quantitative information which is usually the abundance of the molecule under study.

1.1.1 Genomics and genetics data are different

Genomics and genetics data are of different types. Genomics data are related mainly to the collection of DNA sequences modeled as linear strings composed of the four nucleotides symbolized by the letters A, C, G and T (bases). These strings are usually obtained following large sequencing efforts under the supervision of academic and private consortia. NextGen sequencing technologies are used to acquire the data and, specialized softwares are used to assemble sequences in one piece in order to complete an entire genome of thousands of megabases long. The final result of these extensive and costly efforts is the establishment of the genome sequence of all living organisms and particularly the human genome. Genomics has been particularly successful these last few years in determining micro-organism genomes such as bacteria and viruses. Genomics is regularly used in academic research and even proposes on-demand service for the medical community to obtain emerging pathological genomes (SRAS, toxic strains of *Escherichia coli* ...) that allow a fast medical response. Genomics aims to attain the technical challenge of obtaining 99.99% accuracy at the sequenced nucleotide level, and completeness and redundancy in the genome sequence of interest. However, the understanding or interpretation of the genome sequence, which means finding genes and their regulatory signals as well as finding their properties collected under the name "annotations", are still the most challenging and expensive tasks.

Genetics data, or genotype data, are related to the sequencing efforts on the human genome, particularly at the individual level. Genetics data record that the status of some nucleotides found at a certain position in the genome are different from one person to another. These base- and position-specific person-to-person variations are known as SNP or Single Nucleotide Polymorphism. When the frequency of the variation in a population is greater than 1%, this variation is considered as a true polymorphism possibly associated with traits (phenotypes) and genetic diseases (mutations). Moreover this information is useful as a genetic biomarker for susceptibilities to multigenic diseases or ancestrality and migration studies.

1.1.2 Transcriptomics data

Transcriptomics data consist in the recording of the relative abundance of transcripts or mature messenger RNAs representing the level of gene expression in cells when submitted to a particular condition. Messenger RNAs are the gene blueprints or recipes for making the proteins which are the working force (enzymes, framework, hormones...) in a cell and allow the cell's adaptation to its fast changing environment. Transcriptomics give a snapshot of the activity of gene expression in response to a certain situation. Generally mRNA abundances are not measured on an absolute scale but on a relative quantitative scale by comparing the level of abundance to a particular reference situation or control. Raw transcriptomics data associated with a certain gene g consist in recording the ratio of the abundances of its specific gene transcript in two biological situations, the test and the control. This ratio reveals if a particular gene is over- or under- expressed in a certain condition relative to the control condition. Moreover, if a set of genes respond together to the environmental stress under study, this is a signature of a possible common regulation control (Figure 2). Furthermore, transcriptomics data are usually organized as for other «omics» data as large tables of n x p. cells with p samples in columns and n genes in rows (Figure 1). A data pre-processing step is necessary before analyzing transcriptomics data. It consists in $\log_2$ intensity ratios transformation, scaling the ratios across different experiments, eliminate outliers. Multivariate analysis tools, particularly PCA, are then used to find a few genes among the thousands that are significantly perturbed by the treatment. The signification level of the perturbation of a particular gene has purely statistical value and means that the level of measured variation in the ratio is not due to pure chance. It is up to the experimentalist to confirm that it is truly the biological factor under study, and not the unavoidable variation coming from technical or analytical origin inherent to the acquisition method, that is responsible for the observations. To estimate this significance level it is absolutely necessary to measure ratios on a certain replicative level, at least three replicates per gene and per situation. ANOVA and multiple testing False Discovering Rate (FDR) estimates are generally used. Further experimental studies are mandatory to confirm transcriptomics observations. Moreover, Pearson correlation coefficient and different linkage clustering methods are used for each gene in order to perform their hierarchical clustering and to group genes with similar behavior or belonging to the same regulation network.

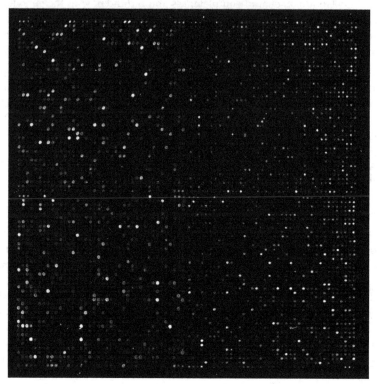

Fig. 2. A picture of a DNA microarray used in high-throughput transcriptomics.

DNA chip of 18 x 18 mm In size containing 6400 yeast gene Specific sequences organized as a matrix in which gene coordinates (x,y) are known. After hybridization with transcripts labeled respectively with green and red fluorochromes from two situations (treated versus untreated), 2 images in red and green fluorescence are recorded and superposed. Spot intensity seen on this image is then mathematically converted to a ratio of relative abundance of gene expression in the two situations under study (DeRisi et al., 1997).

1.1.3 Proteomics and metabolomics data

Proteomics and metabolomics data consist in measuring absolute or relative abundances of proteins and metabolites in the organism, tissue or cells after their proper biochemical extraction. These two fundamental and different classes of molecules are important for preserving cell integrity and reactivity to environment changes. These molecules are generally recognized and their abundances measured by mass spectrometry technologies after a liquid (HPLC) or gas (GC) chromatographic separation is performed to lower the high complexity level of analytes in the sample under study. Proteins have the large size of a few thousands of atoms and weigh a few thousands of Daltons (1 Dalton is the mass of a hydrogen atom) in mass, contrary to metabolites that are smaller molecules in size and mass (less than 1000 Daltons). Mass spectrometers are the perfect analytical tool to separate physically ionized analytes by their mass-to-charge ratio (m/z) and are able to record their abundance (peak intensity). Mass spectrometry data are represented graphically by a spectrum containing abundances versus m/z ratios or by a table or a peak list with two columns containing m/z and abundances after performing a de-isotopic reduction step and a noise filtration step.

Because of the large size of protein molecules, entire proteins should be cut in small pieces, called peptides, of 10-15 amino acids by using a protease enzyme trypsin. These peptides then have the right size to be analyzed directly by mass spectrometers. Peptide abundances are recorded. and their sequences even identified by collision-induced fragmentation (CID) breaking their peptide bonds, which some mass spectrometers instruments can perform (Triple Quadrupole mass spectrometrer in tandem, MALDI TOF TOF, Ion traps).

Raw data from metabolomics and proteomics studies originating from mass spectrometry techniques have the same basic contents. However, contrary to previous «omics», analytes are first separated by a chromatographic step and one analyte is characterized by its unique retention time (rt) on the separation device, its mass-to-charge ratio (m/z) and its abundance (a). This triad (rt - m/z - a) is a characteristic of the analyte that is measured accurately and found in the final «omics» data matrix n x p. Because of the separation step, multiple chromatography experiments should be normalized on both the scale of abundance and the scale of retention time to be further compared. A relevant multiple alignment of the chromatographic separations of different p samples is necessary and is performed by using sophisticated methods and models (Listgarten & Emili, 2005). This alignment step consists in recognizing which analyte is recorded in a given retention time bin and in a given m/z bin. Analytes found in common in the chosen bin are by definition merged in intensity and considered to be the same analyte. The same m/z analyte is recorded across multiple chromatographic steps and should be eluted at the same rt with some degree of tolerance both on rt (a few minutes) and on m/z (a few 0.1 m/z). The rows in the prote- and metabol-«omics» final matrix n x p contain the. proxy "m/z_rt," or "feature" and on the columns are the samples where the analytes come from. The cell content of this matrix record the abundance. "m/z_rt" is a set of analytes which have the same m/z with the same retention time rt, hopefully only one. Data can also be visualized as a 3D matrix with 3 dimensions: rt, m/z and abundances (Figure 3). For convenience it is the "m/z_rt" versus the 2D sample matrix which is further used in EDA for sample comparisons. The absolute value of intensity of the m/z analyte with retention rt corresponds to the mass spectrometry response given by its detector (cps).

1.2 Technologies generating «omics» data, their sizes and their formats

1.2.1 Genetics data format

Genome-wide studies using genetics data consist in recording the status of a particular DNA position or genotype in the genome called SNP or Single Nucleotide Polymorphism among few thousand of genes for a certain number of samples. The SNP status is obtained by accurately sequencing genomic DNA and recording its sequence in databases such as Genbank (www.ncbi.nlm.nih.gov/genbank). The SNP status is then coded by a simple number, 0, 1, 2, according to the nature of the nucleotide found at the genome's particular

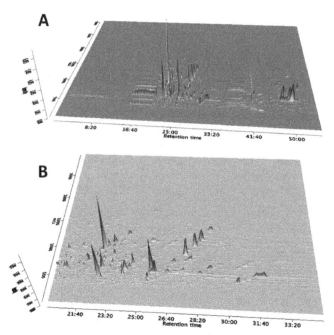

Fig. 3. A 3D representation of a mass spectrum of a liquid chromatographic separation in LC-MS typical analysis of proteomics and metabolomics data.

(A) Urinary native peptides without noise filtration and (B) with noise filtration are shown on a smaller time scale (20 to 34 minutes). These spectra were obtained using MZmine 2.1 with raw data converted first to the mzxml format (Pluskal et al., 2010).

position. It is not rare for the n x p matrix used in genetics data to have for dimension n=500000 SNP positions recorded for p=1000 individuals grouped according to ethnical, geographical or disease status. SNP positions, sequence, type and frequencies are maintained and accessible on different websites such as dbSNP (www.ncbi .nlm.nih.gov/projects/SNP), the International HapMap project (hapmap.ncbi.nlm.nih.gov), the SNP consortium (snp.cshl.org), the Human Gene Mutation Database or HGMD (www.hgmd.org), the 1000 Genomes project (www.1000genomes.org), the Pharmacogenomics database PharmGKB (www.pharmgkb.org) and the genotype-phenotype association database. GWAS Central (www.gwascentral.org). This information is particularly relevant in order to attempt SNP associations to disease status or health conditions. In recent human genetic studies, genotype data have been harvested, consisting in collecting for a few thousand human samples of different classes (ethnic groups, disease status groups, and so on) all the SNP profiles for particular genes (or even better all the genome). Algorithms such as EIGENSOFT suite is used to find statistically acceptable genotype-phenotype associations (Novembre & Stephens, 2008; Reich et al, 2008). The suite contains the EIGENSTRAT tool which is able to detect and correct for population bias of allele frequency, also called stratification, and suggests where the maximum variability resides among the population. PCA was demonstrated as a. valuable tool for detecting population substructure and correcting for stratification representing allele frequency differences originating from ancestry between the considered population before associating SNPs profile and disease status (Price et al., 2006). These studies were recently published for making qualified inferences about human migration and history.

1.2.2 Transcriptomics data formats

In order to analyze in parallel the large population of mRNAs or transcriptomes that a cell is expressing, a high-throughput screening method called DNA microarrays is used today. These DNA chips, some of which are commercially available (ex: Affymetrix), contain imprinted on their glass surface, as individualized spots,. thousands of short nucleic acid sequences specific of genes and organized in matrices to facilitate their location. (Figure 2) Pangenomic DNA chips contain sequences representing ALL the genes known today for a certain species (a few tens of thousands). These chips are hybridized with equal quantity of mRNA or complementary DNA copies of the mRNA prepared from control and treated samples, including fluorescent red (treated) and green (control) nucleotide analogs, in order to keep track of the sample origin. After subsequent washing steps, green and red fluorescence signals present on the chip are measured and the red-to-green ratio is calculated for each gene. The colors of the spots are from red (treated) to green (control) indicating over- and under- abundance of gene expression in the treated condition. A yellow color indicates an equal abundance of gene expression (no effect of condition) and a black spot indicates absence of gene expression in both conditions. Major free-access transcriptomics databases are the Stanford microarray database (smd.stanford.edu) and the NCBI GEO omnibus (www.ncbi.nlm.nih.gov/geo). The size of these arrays depends on the gene population under study. It is not rare to study transcriptomics on n= 7000 genes (yeast) or more on pangenomic arrays n = 20000 – 30000 (Arabidopsis, humans, mice …). The number of DNA microarrays p is generally of the order of a few tens to a few hundreds, taking into account experimental replicates.

Alternative techniques exist to study gene expression, but they are not applied on a large- or genomic-wide scale as DNA microarrays,. and they are used in order to confirm hypotheses given by these later experiments. Among them, the technique using qRT-PCR or quantitative Reverse Transcription Polymerise Chain Reaction or its semi-high throughput variant called microfluidic cards (AppliedBiosystems) allow to quantify gene expression focused on 384 selected genes in one sample.

1.2.3 Proteomics and metabolomics data formats

Numerous mass spectrometry technologies are available today to perform proteomics and metabolomics analyses in specialized laboratories. These «omics» have not yet attained the mature status and the standardization level that transcriptomics has now attained, particularly at the level of data acquisition, data storage and sharing, as well as data analysis. However, some consortia, such as the human proteomic organization HUPO (www.hupo.org) and PeptidesAtlas (www.peptidesatlas.org), are spending a considerable amount of efforts and money to find standardization rules. One of the main difficulties in working with these «omics» data resides in maintaining intra- and inter-laboratory reproducibility. The second difficulty is that few mass spectrometers associated with the chromatographic separation devices are able to record a quantitative signal that is directly proportional to analyte abundance. Semi quantitative data are generally obtained with matrix-assisted laser desorption ionization (MALDI) and quantitative signal is better obtained with electrospray ionization (ESI) methods. The use of relevant working strategies is necessary to lower technical and analytical variabilities, and this is also accomplished through the use of numerous replicates and internal standards with known or predictive mass spectrometry behaviors. The third difficulty is inherent to the commercially available instruments for which data acquisition and processing use computational tools and proprietary data formats. There are, however, a few format converters that are accessible, among them OpenMS (open-ms.sourceforge.net) and TransProteomicPipeline (tools.proteomecenter.org). These techniques are used extensively with the aim of detecting and quantifying biomarkers or molecular signals. specific to drug toxicity, disease status and progression, sample classification, and metabolite pathways analysis. The size of proteomics and metabolomics matrices depends on the accuracy level measured on the analytes mass and the range of mass under study. n varies from a. few hundreds to a few tens of thousands of m/z analytes and the p dimension of experiments or samples is largely dependent on the biological question. (a few tens).

Alternative techniques to confirm proteomic expression use protein chips and immunoprecipitation techniques that are antibody dependent. Another mass spectrometry technique, which is Liquid Chromatography coupled to Selected Reaction Monitoring (LC-SRM), is also used to confirm the level of expression of a particular analyte focusing on its physico-chemical properties as well as its chemical structure. In this case a high specificity and sensibility are generally obtained because the mass spectrometer records the signal related to both occurrences of the presence of one analyte mass (precursor ion) and the presence of one of its specific fragments obtained by collision-induced dissociation (CID).

2. Exploratory data analysis with PCA of «omics» data

2.1 The principles of PCA in «omics»

The original. n x p matrix (Figure 1) or its transposed p x n (Figure 4) contains raw data with n generally much larger than p. The data should be preprocessed carefully. according to the nature and accuracy of the data. In the «omics», particularly proteomics and metabolomics, autoscaling and Pareto normalizations are the most used (Van der Berg et al, 2006). Autoscaling is the process of rendering each variable of the data (the «omics» item) on the same scale with a mean of 0 and a standard deviation of 1. PCA consists in reducing the normalized n x p matrix to two smaller matrices, an S score matrix and an L loading matrix. The product of scores S and the transposed loadings L' matrix plus a residual matrix R gives the original n x p matrix X according to the formula $X = S*L' + R$. PCs are the dimension (d) kept for S and L matrices and numbered PC_1, PC_2, PC_3.... according to the largest variance they capture. PC_1 captures most of the variability in the data followed by PC_2 and PC_3 (Figure 4). PCA helped originally to detect outliers. PCA axis capture the largest amount of variability in the data that scientists in the «omics» fields want to interpret and to relate to biological, environmental, demographic and technological factors (variability in replicates). Therefore variance in the higher PCs is often due to experimental noise, so plotting data on the first two to three PCs not only simplifies interpretation of the data but also reduces the noise. The scoring plot displays the relationship existing between samples, meaning that two similar samples will be distantly close in the PC space. Furthermore, the loading plot displays the relationship between the «omics» items (genes, peptides, proteins, metabolites, SNPs). A strong correlation between items will be expressed by a linear arrangement of these points in the loading matrix. Moreover PCA biplots representing score and loading scatter plots superposed together are useful to detect the importance of particular loadings («omics» measured items) responsible for separating these sample clusters.

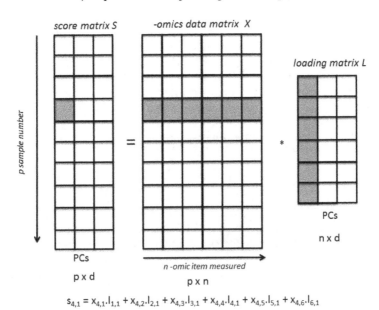

$$s_{4,1} = x_{4,1}{\cdot}l_{1,1} + x_{4,2}{\cdot}l_{2,1} + x_{4,3}{\cdot}l_{3,1} + x_{4,4}{\cdot}l_{4,1} + x_{4,5}{\cdot}l_{5,1} + x_{4,6}{\cdot}l_{6,1}$$

Fig. 4. Relationship between the X «omics» data matrix X, the S score matrix and the L loading matrix in principal component analysis.

Here the original «omics» matrix has been transposed for convenience, with n=6 being the number of omic items experimentally measured and p = 10 being the number of samples considered. d=3 is the number of PCs retained in the model, facilitating its graphical exploration. The highlighted row in X and column in L show what is required to generate a PC_1 score for sample 4.

PCA could reveal main patterns in the data but can detect some systematic non-biologically related or unwanted biologically related bias defined as batch effects. The existence of batch effects in «omics» data is being more and more recognized to frequently misguide biological interpretations. A large number of softwares can calculate these S and L matrices for large data sets. A PLS toolbox from Eigenvector research (www.eigenvector.com) running under MATLAB (www.matworks.com) contains representative 2D or 3D graphics of PCs space. Moreover, statistical indexes such as Q residues allow to estimate the disagreement behavior of some variables (samples) in the model, and Hotelling's T2 indexes measures the multivariate distance of each observation from the center of the dataset. R (cran.r-project.org) contains, in the statistical package included in the basic R installation, the prompt() function, which performs PCA on the command line. Particular «omics» -specific softwares containing sophisticated normalization, statistics and graphic options for proteomics and metabolomics data are available, such as DAnteR (Poplitiya et al., 2009) and MarkerView (www.absciex.com).

2.2 Interactive graphic exploratory data analysis with Ggobi and rggobi

Scatter plots are still the simplest and most effective forms of exploratory analyses of data but are limited to a pairwise comparison with just two samples in any one scatterplot diagram. Ggobi (www.ggobi.org) and rggobi, an alternative of ggobi with R GUI interface (www.ggobi.org/rggobi), are free tools that allow doing a scatter plot matrix with some limitation, as they graphically display a small number of explicative variables (less than 10). Ggobi and rggobi have an effective way of reducing a large multivariate data matrix into a simpler matrix with a much smaller number of variables called principal component or PCs without losing important information within the data. Moreover, this PC space is graphically displayed dynamically as a Grand Tour or 2D tour. Moreover, samples can be specifically colored or glyphed by using a "brushing" tool according to their belonging to some factors or categorical explicative variables (patient status, sample group, and so on…). Moreover, unavailable measurements (NA) are managed by Ggobi by using simple value replacements (mean, median, random) as well as sophisticated multivariate distribution modeling (Cook & Swayne, 2007).

2.3 PCA for «omics» data

2.3.1 PCA for genetics data

PCA was used almost 30 years ago by Cavalli-Sforza L.L in population genetics studies to produce maps summarizing human genetic variation across geographic regions (Menozzi et al., 1978). PCA is used also in genotype-phenotype association studies in order to reveal language, ethnic or disease status patterns (Figure 5). Recently it has been shown that these studies are difficult to model with PCA alone because of the existence of numerous unmeasured variables having strong effects on the observed patterns (Reich et al., 2008; Novembre & Stephens, 2008). When analyzing spatial data in particular, PCA produces highly structured results relating to sinusoidal functions of increasing frequency with PC numbers and are sensitive to population structure, including distribution of sampling locations . This observation has also been seen in climatology. However PCA can reveal

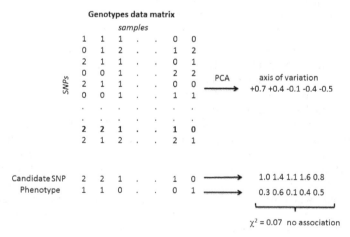

Fig. 5. The EIGENSTRAT algorithm from the EIGENSOFT suite.

some patterns on the data, and reliable predictions require further genetic analysis and integration with other sources of information from archeology, anthropology, epidemiology, linguistic and geography (François et al., 2010).

Genotype data consists of a n x p matrix where p individuals are recorded for their n SNPs. in their genomes. PCA is applied to infer continuous axes of genetic variation. A single axis of variation is indicated here. A genotype at a candidate SNP and phenotype are continuously adjusted by amounts attributable to ancestry along each axis. A χ^2 show here no significant association for this particular SNP (Price et al., 2006).

2.3.2 PCA for transcriptomics data

Gene expression array technology has reached the stage of being routinely used to study clinical samples in search of diagnostic and prognostic biomarkers. Due to the nature of array experiments, the number of "null-hypotheses" to test, one for each gene, can be huge (a few tens of thousands). Multiple testing corrections are often necessary in order to. screen non-informative genes and reduce the number of null-hypotheses. One of the commonly used methods for multiple testing control is to calculate the false discovery rate (FDR) which is the ratio of the number of false rejections among the total number of rejections. FDR adjustment on raw p-values is effective in controlling false positives but is known to reduce the ability to detect true differentially expressed genes.

In transcriptomics studies, PCA is often used for the location of genes relative to each other in a reduced experiment space. Genes are plotted with respect to the two orthogonal linear combinations of experiments that contain the most variance (Lu et al., 2011). Transcriptomics also use other multivariate tools for classification and clustering (Tibshirani et al., 2002). A very fast and effective classification strategy is linear discriminant analysis. In classification problems there are positive training examples that are known members of the class under. study and negative training examples that are examples known not to be members of the class. The test examples are compared to both sets of training examples, and the determination of which set is most similar to the test case is established. In this process the test example is "classified" based on training examples. Clustering is a commonly used categorizing technique in many scientific areas using K-means grouping technique. Using this approach the user can cluster data based on some specified metric into a given number of clusters. Users can cluster arrays or genes as desired into a pre-specified number of clusters. The algorithm has a randomized starting point so results may vary from run to run.

2.3.3 PCA for proteomic and peptidomic data

2.3.3.1 Urinary peptides and biomarker discovery study

PCA was used in order to distinguish urine samples containing or not pseudo or artificial spiked-in analytes or pseudo biomarkers (Benkali et al., 2008). The objectives were to analyze variations in the data and distinguish their sources. These variations could arise from (a) experimental variations due to changes in the instrument or experimental conditions, (b) real variations but of no interest in the primary objective, such as male versus female subjects, drug treatments, metabolites of a therapeutic agent... and (c) relevant differences that reflect changes in the system under study (spiked-in or not spiked-in). The experiment consisted in using human urines from 20 healthy volunteers splitted in two groups of ten, one which was spiked-in with few synthetic peptides at a certain variable concentration and the other without. Urines were processed using the same peptide extraction solid phase extraction (SPE) protocol, by the same experimentalist, and peptide compositions were recorded by off-line nanoLC-MS MALDI TOF/TOF. Data were processed with MarkerView software version 1.2 (www.absciex.com). PCA preprocessing consisted in using Pareto scaling without weighing and no autoscaling because Pareto scaling is known to reduce but not completely eliminate the significance of intensity, which is appropriate for MS because larger peaks are generally more reliable and all variables are equivalent. Different scaling methods are worth trying because they can reveal different features in the data with peak finding options and Pareto normalization (Van der Berg et al., 2006).

More than 5000 features (or m/z analytes) were retained from which respective abundances were observed. The n x p matrix contains n= 5000 and p = 20 samples. Scores and loading on PCs were calculated with 3 PCs capturing 80.4% of total data variability. Figure 6 shows PC_1 (70.6%) versus PC_2 (7.4%) (Figure 6A), as well as. PC_1 (70.6%) versus PC_3 (2.4%) (not shown). Sample points in the scoring scatterplot were colored according to. their group assignment before analysis (unsupervised). PCs scores on the PC_1-PC_2 projection axis allowed us to define the A9 sample as an outlier behaving as an unspiked B group sample (labeling tube error perhaps). We had to discard this sample for the rest of the analysis. This analysis was carried out on samples blinded to categorical label (spiked and unspiked) and the coloring of samples on the graphic was only carried out after the PCA. Spiked samples (A samples) are in red color and unspiked samples in blue color (B samples). The high positive value of loadings (green points) in the PC_1 and PC_2 axes are associated with features (or m/z analytes) most responsible to discriminate the two sample groups. The relative abundance of the spiked analyte of m/z = 1296.69 and its two [13]C. isotopically stable labeled variants, 1297.69 and 1298.69, is shown in the spiked group (A samples, red points) and in the unspiked group (B samples, blue points). Moreover they tend to lie close to straight lines that pass through the origin in the loading plots (Figure 6A). These points (green points) are correlated because they are all the isotopic forms of the same spiked compound. The same is observed for other spiked analytes (904.46, 1570.67, 2098.09 and 2465.21). Finally superposed mass spectra from 20 samples of both groups show the relative abundance of analytes (panel B and insert focused on m/z = 1296.69 analyte and its natural ^{13}C isotopes).

Variables with large positive PC_1 loadings are mainly in group A (spiked samples) and absent or at lower intensities in group B. PC_1 separates both groups but PC_2 seems to separate both groups A and B in two half-groups (Figure 6A). What is the nature of the variation captured by PC_2 where some loadings (900.40, 1083.54 and 1299.65) give high positive PC_2 values and negative PC_1 values ?. The examination of Figure 7 shows that these analytes show a progessive increase in their intensity with a gradient following the order of their analysis in the instrument. The data were acquired in the order they are displayed (from left to right) and group A members were acquired before members of group B, which introduces a bias or a batch effect in these data. To avoid this effect, the samples should be acquired in a random order,. with group members mixed.

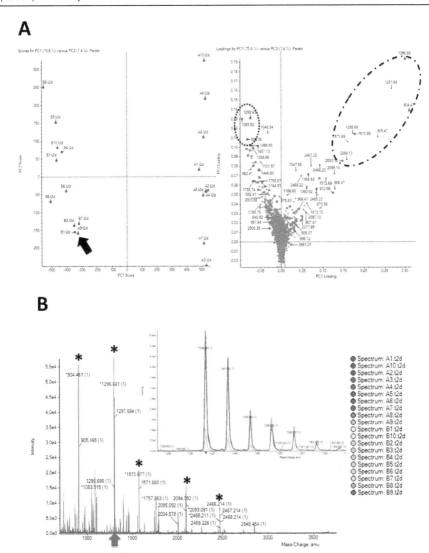

Fig. 6. PCA of 20 urine samples spiked or not spiked with synthetic peptides (pseudo biomarkers).

(A) Scores and loadings plots of PC₁ and PC₂ axes show a good separation of group A (spiked,. colored in red) from group B (not spiked, colored in blue). The A9 sample (black arrow) is an outlier and behaves as a B member group. It should be removed from the analysis. Loading plots show the 5000 analytes (green points) from which the majority are not contributing to the variability (0,0). Some analytes contribute to the large positive variation in the PC₁ axis (spiked peptides) and to the positive PC₂ (bias effect). (B) Superposition of the 20 spectra of urine samples after their alignment. with. a symbol (*) indicating the m/z of spiked peptides. The insert corresponds to the enlargement of the spectra located at the red arrow in the spectra, showing the abundance of. the 1296.69. and their ¹³C isotopes among the 20 samples, particularly in the A group.

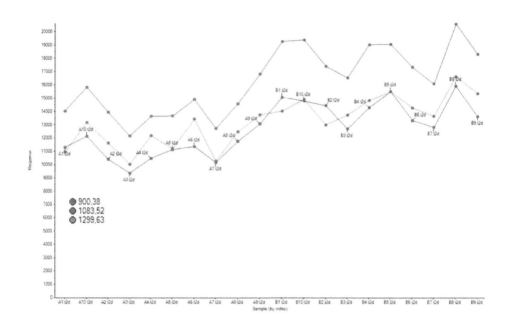

Fig. 7. A variation. recognized in the PC_2 axis probably due to a batch effect in the sample processing. Analytes. 900.38, 1083.52 and 1299.63. responsible for the positive value of scores in the PC_2 axis (Figure 6, panel A) see their intensity increase slightly during their acquiring time (from left to right), signaling a probable batch effect.

2.3.3.2 Microbial identification and classification by MALDI-TOF MS and PCA

MALDI-TOF MS mass spectrometry has recently been used as a technique to record. abundance of proteins extracted from different phyla of bacteria with the aim of finding phylum-specific patterns and use them to classify or recognize these bacteria in a minimum culture time. Sauer, S. has pioneered the technique of rapid extraction of proteins from alcohol, strong acid treatment or direct transfer from single colonies of bacteria, with or without the need to cultivate them (Freiwald & Sauer, 2009). Ethanol/Formic acid extraction of proteins of two clones of each. 6 bacteria strains, *Klebsiela pneumonia* (KP), *Acinetobacter baumanii*, (AB), *Lactobacillus plantarum* (LP), *Pseudomonas aeruginosa* (PA), *Escherichia coli* MG (MG), *Bacillus subtillis* (BS) were prepared. Mass spectra of proteins were recorded in five analytical replicates in the range 4000 to 12000 Daltons (Figure 8A). Major extracted proteins come from abundant ribosomal proteins. The natural variants in their amino acid sequence are responsible for the differences of masses in the peaks observed in the spectra, and their abundance is characteristic of the bacteria. Moreover, 6 bacteria clones (X1 to X6) were blindly analyzed in triplicate. PCA was used in order to distinguish the axes of greater variability in the data.

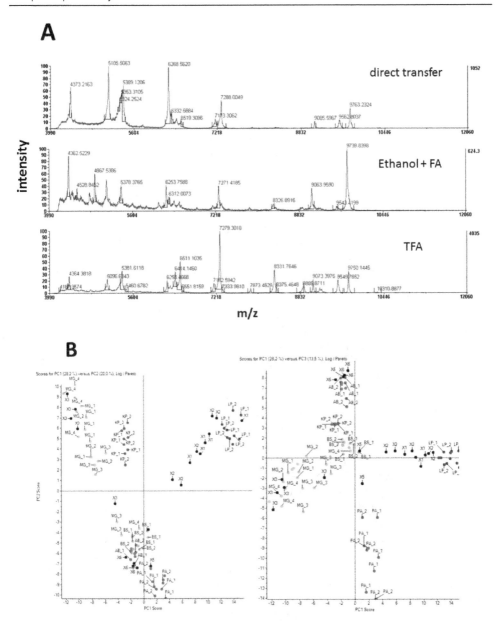

Fig. 8. MALDI TOF MS spectra of protein extracts of 6 bacteria strains and PCA.

(A) MS spectra of *Escherichia coli* MG clone proteins extracted with 3 different methods, direct transfer, ethanol/Formic acid (FA). and Trifluoroacetic acid(TFA). Peak abundances. may vary according to the extraction methods, from bacteria strains and from analytical variability. (B) Scores of PC_1 versus PC_2 and PC_1 versus PC_3 of the full 6 bacteria dataset,. including unknown X1 to X6 samples. (black points). (unpublished Maynard & Gastinel, 2009).

For this analysis, PC_1 takes 28.2% of variability, PC_2 20% and PC_3 15%, for a total of 63.2% captured in the model (Figure 8B). The PC_1 axis separates LP and PA bacteria strain. from. KP, MG, BS and AB strains. The PC_2 and PC_3 axes separate. KP and LP from BS , AB and PA. but not so well for the MG strain. Moreover, X samples are 100% separated in their correct respective clusters. The broad distribution of samples in the score plot is probably due to the relatively poor analytical reproducibility in the ionization of. MALDI TOF MS analysis. From the loading score plots (not shown) few proteins of particular m/z responsible for this bacteria strain separation have been recognized in protein databases by their annotation. Among them, the 5169 Daltons protein is attributed to the 50S ribosomal protein annotated as L34 of *Acinetobater baumanii*.

2.3.4 PCA for metabolomics data

Humic acids are one of the major chemical components of humic substances, which are the major organic constituents of soil (humus), peat, coal, many upland streams, dystrophic lakes, and ocean water. They are produced by biodegradation of dead organic matter. They are not a single acid; rather they are a complex mixture of many different acids containing carboxyl and phenolate groups so that the mixture behaves functionally as a dibasic acid or, occasionally, as a tribasic acid. Humic acids can form complexes with ions that are commonly found in the environment, creating humic colloids. Humic and fulvic acids (fulvic acids are humic acids of lower molecular weight and higher oxygen content than other humic acids) are commonly used as a soil supplement in agriculture, and less commonly as a human nutritional supplement. Humic and fulvic acids are considered as soil bioindicators and reflect an equilibrium between living organic and non-organic matters.

Mass spectrometry has been used to estimate signature analytes and patterns specific to some soils (Mugo & Bottaro, 2004). Fulvic acids were prepared from a soil using different extraction protocols resulting in 5 samples,. H1, H1H2, EVM1, EVM2 and EAA. Are these extraction protocols similar and which analytes are they extracting more efficiently? MALDI MS spectra from 150 to 1500 m/z range were recorded in the presence of the MALDI matrix alpha-cyano-4-hydroxycinammic acid (CHCA). Normalization of intensities were done with the 379 m/z analyte in common to these samples, and Pareto scaling was chosen during the alignment process performed by MarkerView. Figure 9A shows that the PCA analysis reveals poor separation of samples with PC_1 explaining 25.8%, PC_2 18.7% and PC_3 14.8%. of variability (a total of 59.3% captured). Samples are not so well separated by the first PC axis, demonstrating the large influence of factors other than soil extraction differences (chemical precipitation, physical precipitation, filtration). Discriminant Analysis associated with PCA (supervised PCA-DA) was attempted to further separate these known 5 groups (Figure 9B). This supervised technique means that it uses class information based on the assigned sample group to improve their separation. Figure 9B shows a dramatic improved separation but this may be based on noise. Peaks which are randomly more intense in one group as compared to another can possibly influence the results, and careful examination of loading plots as well as analyte profiles across the samples is necessary to avoid batch effects. This analysis is also affected by samples incorrectly assigned to wrong group and outliers.

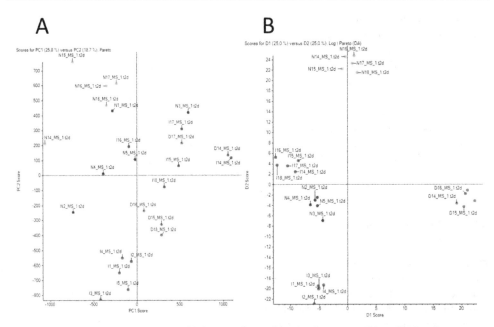

Fig. 9. Mass spectrometric study of fulvic acids profiles in the range 150 to 1000 m/z present in 5 sample preparations and analyzed in five analytical replicates by PCA.

(A) Unsupervised PCA and (B) PCA-DA or supervised PCA with group information included before reducing data. Only the first and second PC axes are shown for the score plots. (unpublished Basly & Gastinel, 2009).

In Figure 9, PCA (A) and PCA-DA (B) show a different pattern of the variability in the dataset. PC_1. axis does not discriminate H1H2 (green) , EVM1 (blue) and EAA (violet) as for. the PC_2 axis. The PC_2 axis discriminates however EVM2 (pale green). and H1 (red). Keeping these 5 different protocols as 5 different classes, PCA-DA (Figure 9B) however discriminates quite well these these 5 preparation analyses in quintuplicates. EVM1 and EAA are still the closest group. Loading score plots reveal which analytes are the most favored in a particular extraction method relative to another.

3. How can PCA help to reveal batch effects in «omics» data?

3.1 What are batch effects?

Batch effect is one overlooked complication with «omics» studies and occurs because high-throughput measurements are affected by multiple factors other than the primary tested biological conditions (Leek et al, 2010; Leek & Storey, 2008). These factors are included in a comprehensive list among which are laboratory conditions, reagents batches, highly trained personnel differences, and hardware maintenance. Batch effect becomes a problem when these conditions vary during the course of an experiment, and it becomes a major problem when the various batch effects are possibly correlated with an outcome of interest and lead to incorrect conclusions (Ransohoff, 2005; Baggerly, et al., 2004). Batch effects are defined as a sub-group of measurements that have qualitatively different behaviors across conditions and are primarily unrelated to the biological or scientific variables under study. Typical batch effect is seen when all samples of a certain group are measured first, and when all samples of a second group are measured next. Batch effect occurs too when a particular batch of reagent (ex: Taq polymerase enzyme for PCR experiments) is used with all samples of the first group, and another reagent batch is used with all samples of the second group. Typical batch effects are also seen when an experimentalist/technician acquires all samples from the first group and a different experimentalist/technician works with the other group or when the instrument's characteristics (example for MALDI mass spectrometry: laser or detector replacements) used to acquire the data have been deeply modified. Data normalization generally does not remove batch effect unless normalization takes into account the study design or takes into account the existence of a batch problem.

3.2 How to find evidence of batch effects

The first step in addressing batch and other technical effects is to develop a thorough and meticulous study plan. Studies with experiments that run over long periods of time, and large-scale, inter-laboratory experiments, are highly susceptible to batch effects. Intra-laboratory experiments spanning several days and several personnel changes are also susceptible to batch effects. Steps necessary to analyze batch effects require different. levels of analysis, according to the recent review of Leek J.T (Leek et al., 2010). What follows are some of the recommended actions:. Performing a hierarchical clustering of samples that assigns a label to the biological variables and to the batch surrogates estimates, such as laboratory and processing time; plotting individual features (gene expression, peptides or metabolites abundances). versus biological variables and batch surrogates using ggobi for example; calculating principal components of the high-throughput data and identifying components that correlate with batch surrogates. If some batch effects are present in the data, artifacts must be estimated directly, using surrogate variable analysis (SVA) (Leek et al., 2007). Recently, the EigenMS algorithm has been developed and. implemented within a pipeline of bioinformatic tools of DanteR in order to correct for technical batch effects in MS proteomics data analysis (Polpitya et al., 2008; Karpievitch et al., 2009). The algorithm uses an SVA approach to estimate systematic residual errors using singular value decomposition taking account primary biological factors and substracting those estimates from raw data in the pre-processing data analysis. The estimated/surrogate variables should be treated as standard covariates in subsequent analyses or adjusted for use with tools such as Combat (Johnson & Li, 2007). After adjustments that. include surrogate variables (at least processing time and date), the data must be reclustered to ensure that the clusters are not still driven by batch effects.

3.3 How to avoid batch effects

Measures and steps must be taken to minimize the probability of confusion between biological and batch effects. High-throughput experiments should be designed to distribute batches and other potential sources of experimental variation across biological groups. PCA of the high-throughput data allows the identification of components that correlate with batch surrogate variables.

Another approach to avoid and prevent batch effects is to record all parameters that are important for the acquisition of the measures and the relevant information related to demographic and grouping factors. The structure of a database under MySQL with attractive web graphic user interface (GUI) should be conceived at the same time as the study design is defined. Such a database was constructed for a mass spectrometry based biomarker discovery project in kidney transplantation in a French national multicenter project. The BiomarkerMSdb database structure contains 6 linked tables: Demographic data, Peptide Extraction Step, Liquid Chromatography Separation, Probot Fractionation, Spectrometry Acquisition and Data Processing. Figure 10 shows the details of the form that the user must complete to record demographic data of patients enrolled in this project. This approach, with an internet interface used to facilitate data exchange between laboratories enrolled in the project, allows to keep track of essential parameters that could interfere with future interpretations of the results. At minimum, analyses should report the processing group, the analysis time of all samples in the study, the personnel involved, along with the biological variables of interest, so that the results can be verified independently. This is called data traceability.

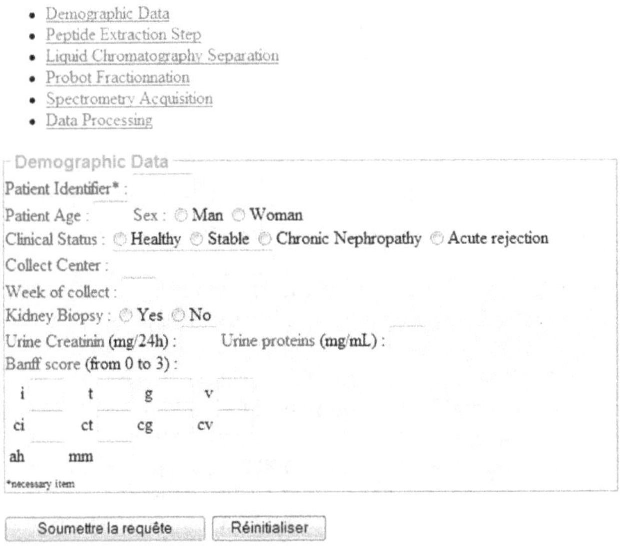

- Demographic Data
- Peptide Extraction Step
- Liquid Chromatography Separation
- Probot Fractionnation
- Spectrometry Acquisition
- Data Processing

Demographic Data

Patient Identifier* :
Patient Age : Sex : Man Woman
Clinical Status : Healthy Stable Chronic Nephropathy Acute rejection
Collect Center :
Week of collect :
Kidney Biopsy : Yes No
Urine Creatinin (mg/24h) : Urine proteins (mg/mL) :
Banff score (from 0 to 3) :
i t g v
ci ct cg cv
ah mm
*necessary item

Soumettre la requête | Réinitialiser

Fig. 10. Extract of the internet form used to interact with BiomarkerMSdb, a relational database under MySQL constructed to record essential parameters involved in a biomarker discovery project using LC-MS strategies. This form is used to fill one of the 6 tables of the database called. "Demographic Data"(unpublished Moulinas & Gastinel, 2011).

4. Conclusion and perspectives

Observational and experimental biology is confronted today with a huge stream of data or "omics" data acquired by sophisticated and largely automated machines. This data is recorded under a digital format that has to be stored safely and shared within the scientific community. One of the challenges of modern biologists is to extract relevant knowledge from this large data. For that purpose, biologists not only should be acquainted with the use of existing multivariate and statistical data mining tools. generally used by meteorologists, economists, publicists and physicists but also should conceive their own specific tools. Among the tools available for multivariate analysis of «omics» data, Principal Component Analysis (PCA) as well as PLS and PCA-DA derivatives have demonstrated their utility in visualizing patterns in the data. These patterns consist sometimes in detecting outliers that spoiled data and that could be removed from them. PCA quantify major sources of variability in the data and allows to show which variables are most responsible for the relationship detected between the samples under study. Moreover, unwanted sources of variability can be revealed as batch effects and partially corrected by surrogate variable analysis (SVA) and EigenMS approaches. However, there are some limitations to using PCA in "omics" data. These limitations result from the large choice of methods of the data pre- and post-processing and the technical difficulty in displaying graphically all the data. Ggobi and rggobi allow to display quite large data using Grand tour and 2D tour, showing dynamic projections with the ability to color and glyph points according to factors (brushing). PCA is an invaluable tool in the preliminary exploration of the data and in filtering or screening them according to noise, outliers and batch effects before using other multivariate tools such as classification and clustering. An appropriate educational program should be pursued in universities in order to expose the theory and praticability of these tools to future biologists.

5. References

Baggerly, K.A., Edmonson, S.R., Morris, J.S. & Coombes, K.R.(2004) High-resolution serum proteomic patterns for ovarian cancer detection. *Endocrine-Related Cancer*, 11, 583-584.

Basly, J.C & Gastinel, L. (2009) Influences of extraction protocols in fulvic acids preparations from soils. *Master 2 Training Report*, GREASE Department, University of Limoges, France.

Benkali, K., Marquet,P.,. Rerolle, J., LeMeur,Y. & Gastinel, L. (2008) A new strategy for faster urine biomarker identification by nano-LC-MALDI-TOF/TOF mass spectrometry. *BMC Genomics*, 14, 9, 541-549.

Cavalli-Sforza,L.L., Menozzi,P. & Piazza,A. (1994) The History of Geography of Human Genes, Princeton University Press, ISBN: 9780691087504, NJ, USA.

Cook, D & Swayne, DF (2007) Interactive and dynamic graphics for data analysis with R and ggobi. Springer, ISBN: 978-0-387-71761-6, NJ, USA.

DeRisi,J.L., Vishwanath, R.I & Brown, P.O. (1997) Exploring the Metabolic and Genetic Control of the Gene Expression on a Genomic Scale. *Science*, 278, 681 – 686.

François, O., Currat,M., Ray, N., Han, E., Excoffier, L.& Novembre, J. (2010) Principal component analysis under population genetic models of range expansion and admixture. *Molecular Biological Evolution*, 27, 6, 1257-1268.

Freiwald, A. & Sauer, S (2009) Phylogenetic classification and identification of bacteria bymass spectrometry", *Nature protocols*, 4, 5, 732-742.

International Human Genome Sequencing Consortium (2001) Initial sequencing and analysis of the human genome, *Nature*, 409,860-921.

Johnson, W.E. & Li, C (2007) Adjusting batch effects in microarray expression data using empirical Bayes methods. *Biostatistics*, 8,1,118-127.

Karpievitch, Y.V., Taverner, T., Adkins, J.N., Callister, S.J., Anderson, G.A., Smith, R.D. & Dabney, A.R. (2009) Normalization of peak intensities in bottom-up MS-based proteomics using singular value decomposition. *Bioinformatics*, 25, 9, 2573-2580.

Leek,J.T., Scharpf, R.B., Corrada –Bravo, H., Simcha, D., Langmead, B., Johnson, W.E., Geman, D., Baggerly, K. & Irizarry,R.A (2010) Tackling the widespread and critical impact of batch effects in high-throughput data. *Nature Genetics*, 11, 733-739.

Leek,J.T. & Storey, D.J (2008) A general framework for multiple testing dependence. *Proceeding of the National Academy of Sciences*, 105, 48, 18718-18723.

Listgarten, J. & Emili, A. (2005) Statistical and Computational methods for comparative proteomic profiling using liquid chromatography-tandem mass spectrometry. *Molecular and Cellular Proteomics*, 4.4, 419-434.

Lu, J., Kerns, R.T., Peddada, S.D. & Bushel, P.R. (2011) Principal component analysis-based filtering improves detection for Affymetrix gene expression. *Nucleic Acids Research*, 2011, 1-8.

Maynard,C. & Gastinel, L. (2010) Phylogenetic relationship between bacteria revealed by MALDI TOF mass spectrometry. *Life Science Undergraduate Training Report*, Microbiology Department, University of Limoges, France.

Menozzi, P., Piazza, A & Cavalli-Sforza,L (1978) Synthetic maps of human gene frequencies in Europeans. *Science*, 201,4358, 786-792.

Mugo, S.M. & Bottaro, C.S. (2004) Characterization of humic substances by matrix-assisted laser desorption/ionization time-of-flight mass spectrometry. *Rapid Communication in. Mass Spectrometry.* 18,20, 2375-2382.

Moulinas, R & Gastinel,L (2011) How to detect batch effect in mass spectrometry analysis – Constitution of BiomarkerMSdb. *Master 1 SVT Report* , University of Limoges, France.

Novembre, J. and Stephens, M. (2008) Interpreting principal component analyses of spatial population genetic variation. *Nature Genetics*, 40, 5, 646-649.

Pluskal, T., Castillo, S., Villar-Briones, A., & Oresic, M. (2010) MZmine 2: Modular framework for processing, visualizing, and analyzing mass spectrometry-based molecular profile data. BMC *Bioinformatics* 11:395-405.

Poplitiya,A.D., Qian, W-J., Jaity, N., Petyuk, V.A., Adkins, J.N., Camp II, D.G., Anderson, G.A. and Smith, R.D (2008) DAnTE: A statistical tool for quantitative analysis of – "omics" data. *Bioinformatics*, 24,13, 1556-1558.

Price, A.L., Patterson, N.J., Plenge, R.M., Weinblatt, M.E., Shadick, N.A. & Reich, D. (2006). Principal components analysis corrects for stratification in genome-wide association studies. *Nature Genetics*, 38,8,904-908.

Ransohoff, D.F. (2005) Lessons from controversy: ovarian cancer screening and serum proteomics. *Journal of The National Cancer Institute*,97, 4, 315-317.

Rao,P.K. & Li, Q (2009) Principal component analysis of proteome dynamics in iron-starved mycobacterium tuberculosis. *J. Proteomics Bioinformatics*, 2,1, 19-31.

Reich, D., Price, A.L. and Patterson (2008) Principal component anaysis of genetic data. *Nature Genetics*, 40, 5, 491-492.

Sah, H.N. & Gharbia, S.E. (2010) *Mass Spectrometry for Microbial Proteomics*, Wiley,ISBN: 978-0-470-68199-2,UK.

Stein, L. (1996) How Perl saved the Human Genome Project, in *The Perl Journal*, 1,2, available from: www.foo.be/docs/tpj/issues/vol1_2/tpj0102-0001.html.

Tibshirani, R., Hastie, T., Narasimhan, B & Chu, G. (2002) Diagnosis of multiple cancer types by shrunken centroids of gene expression, *Proceedings of the National Academy of Sciences*, USA. 99, 10, 6567-6572.

Van der Berg, R., Hoefsloot, H.C.J., Westerhuis, J.A., Smilde, A.K. & Ven der Werf, M.J(2006) Centering, scaling and transformations: improving the biological information content of metabolomics data. *BMC Genomics*, 7, 142-156.

Venter, J.G. et al, (2001) The sequence of the Human Genome, *Science*, 291, 1305-1351.

2

Kernel Methods for Dimensionality Reduction Applied to the «Omics» Data

Ferran Reverter, Esteban Vegas and Josep M. Oller

Department of Statistics, University of Barcelona

Spain

1. Introduction

Microarray technology has been advanced to the point at which the simultaneous monitoring of gene expression on a genome scale is now possible. Microarray experiments often aim to identify individual genes that are differentially expressed under distinct conditions, such as between two or more phenotypes, cell lines, under different treatment types or diseased and healthy subjects. Such experiments may be the first step towards inferring gene function and constructing gene networks in systems biology.

The term "gene expression profile" refers to the gene expression values on all arrays for a given gene in different groups of arrays. Frequently, a summary statistic of the gene expression values, such as the mean or the median, is also reported. Dot plots of the gene expression measurements in subsets of arrays, and line plots of the summaries of gene expression measurements are the most common plots used to display gene expression data (See for example Chambers (1983) and references therein).

An ever increasing number of techniques are being applied to detect genes which have similar expression profiles from microarray experiments. Techniques such clustering (Eisen et al. (1998)), self organization map (Tamayo et al. (1999)) have been applied to the analysis of gene expression data. Also we can found several applications on microarray analysis based on distinct machine learning methods such as Gaussian processes (Chu et al. (2005); Zhao & Cheung (2007)), Boosting (Dettling (2004)) and Random Forest (Diaz (2006)). It is useful to find gene/sample clusters with similar gene expression patterns for interpreting the microarray data.

However, due to the large number of genes involved it might be more effective to display these data on a low dimensional plot. Recently, several authors have explored dimension reduction techniques. Alter et al. (2000) analyzed microarray data using singular value decomposition (SVD), Fellenberg et al. (2001) used correspondence analysis to visualize genes and tissues, Pittelkow & Wilson (2003) and Park et al. (2008) used several variants of biplot methods as a visualization tool for the analysis of microarray data. Visualizing gene expression may facilitate the identification of genes with similar expression patterns.

Principal component analysis has a very long history and is known to very powerful for the linear case. However, the sample space that many research problems are facing, especially the sample space of microarray data, are considered nonlinear in nature. One reason might be that the interaction of the genes are not completely understood. Many biological pathways are still beyond human comprehension. It is then quite naive to assume that the genes should be connected in a linear fashion. Following this line of thought, research on nonlinear dimensionality reduction for microarray gene expression data has increased (Zhenqiu et al. (2005), Xuehua & Lan (2009) and references therein). Finding methods that can handle such data is of great importance if as much information as possible is to be gleaned.

Kernel representation offers an alternative to nonlinear functions by projecting the data into a high-dimensional feature space, which increases the computational power of linear learning machines, (see for instance Shawe-Taylor & Cristianini (2004); Scholkopf & Smola (2002)).

Kernel methods enable us to construct different nonlinear versions of any algorithm which can be expressed solely in terms of dot products, known as the kernel trick. Thus, kernel algorithms avoid the explicit usage of the input variables in the statistical learning task. Kernel machines can be used to implement several learning algorithms but they usually act as a black-box with respect to the input variables. This could be a drawback in biplot displays in which we pursue the simultaneous representation of samples and input variables.

In this work we develop a procedure for enrich the interpretability of Kernel PCA by adding in the plot the representation of input variables. We used the radial basis kernel (Gaussian kernel) in our implementation however, the procedure we have introduced is also applicable in cases that may be more appropriated to use any other smooth kernel, for example the Linear kernel which supplies standard PCA analysis. In particular, for each input variable (gene) we can represent locally the direction of maximum variation of the gene expression. As we describe below, our implementation enables us to extract the nonlinear features without discarding the simultaneous display of input variables (genes) and samples (microarrays).

2. Kernel PCA methodology

KPCA is a nonlinear equivalent of classical PCA that uses methods inspired by statistical learning theory. We describe shortly the KPCA method from Scholkopf et al. (1998).

Given a set of observations $\mathbf{x}_i \in \mathbb{R}^n$, $i = 1, \ldots, m$. Let us consider a dot product space F related to the input space by a map $\phi : \mathbb{R}^n \to F$ which is possibly nonlinear. The feature space F could have an arbitrarily large, and possibly infinite, dimension. Hereafter upper case characters are used for elements of F, while lower case characters denote elements of $\mathbb{R}^n$. We assume that we are dealing with centered data $\sum_{i=1}^{m} \phi(\mathbf{x}_i) = 0$.

In F the covariance matrix takes the form

$$\mathbf{C} = \frac{1}{m} \sum_{j=1}^{m} \phi(\mathbf{x}_j)\phi(\mathbf{x}_j)^\mathsf{T}.$$

We have to find eigenvalues $\lambda \geq 0$ and nonzero eigenvectors $\mathbf{V} \in F \backslash \{0\}$ satisfying

$$\mathbf{CV} = \lambda\mathbf{V}.$$

As is well known all solutions $\mathbf{V}$ with $\lambda \neq 0$ lie in the span of $\{\phi(\mathbf{x}_i)\}_{i=1}^{m}$. This has two consequences: first we may instead consider the set of equations

$$\langle \phi(\mathbf{x}_k), \mathbf{CV} \rangle = \lambda \langle \phi(\mathbf{x}_k), \mathbf{V} \rangle , \tag{1}$$

for all $k = 1, \ldots, m$, and second there exist coefficients α_i, $i = 1, \ldots, m$ such that

$$\mathbf{V} = \sum_{i=1}^{m} \alpha_i \phi(\mathbf{x}_i). \tag{2}$$

Combining (1) and (2) we get the dual representation of the eigenvalue problem

$$\frac{1}{m} \sum_{i=1}^{m} \alpha_i \left\langle \phi(\mathbf{x}_k), \sum_{j=1}^{m} \phi(\mathbf{x}_j) \left\langle \phi(\mathbf{x}_j), \phi(\mathbf{x}_i) \right\rangle \right\rangle = \lambda \sum_{i=1}^{m} \alpha_i \left\langle \phi(\mathbf{x}_k), \phi(\mathbf{x}_i) \right\rangle ,$$

for all $k = 1, \ldots m$. Defining a $m \times m$ matrix K by $K_{ij} := \left\langle \phi(\mathbf{x}_i), \phi(\mathbf{x}_j) \right\rangle$, this reads

$$K^2 \alpha = m\lambda K \alpha, \tag{3}$$

where α denotes the column vector with entries $\alpha_1, \ldots, \alpha_m$. To find the solutions of (3), we solve the dual eigenvalue problem

$$K\alpha = m\lambda \alpha, \tag{4}$$

for nonzero eigenvalues. It can be shown that this yields all solutions of (3) that are of interest for us. Let $\lambda_1 \geq \lambda_2 \geq \cdots \geq \lambda_m$ the eigenvalues of K and $\alpha^1, \ldots, \alpha^m$ the corresponding set of eigenvectors, with λ_r being the last nonzero eigenvalue. We normalize $\alpha^1, \ldots, \alpha^r$ by requiring that the corresponding vectors in F be normalized $\left\langle \mathbf{V}^k, \mathbf{V}^k \right\rangle = 1$, for all $k = 1, \ldots, r$. Taking into account (2) and (4), we may rewrite the normalization condition for $\alpha^1, \ldots, \alpha^r$ in this way

$$1 = \sum_{i,j}^{m} \alpha_i^k \alpha_j^k \left\langle \phi(\mathbf{x}_i), \phi(\mathbf{x}_j) \right\rangle = \sum_{i,j}^{m} \alpha_i^k \alpha_j^k K_{ij} = \left\langle \alpha^k, K\alpha^k \right\rangle = \lambda_k \left\langle \alpha^k, \alpha^k \right\rangle . \tag{5}$$

For the purpose of principal component extraction, we need to compute the projections onto the eigenvectors $\mathbf{V}^k$ in F, $k = 1, \ldots, r$. Let $\mathbf{y}$ be a test point, with an image $\phi(\mathbf{y})$ in F. Then

$$\left\langle \mathbf{V}^k, \phi(\mathbf{y}) \right\rangle = \sum_{i=1}^{m} \alpha_i^k \left\langle \phi(\mathbf{x}_i), \phi(\mathbf{y}) \right\rangle , \tag{6}$$

are the nonlinear principal component corresponding to ϕ.

2.1 Centering in feature space

For the sake of simplicity, we have made the assumption that the observations are centered. This is easy to achieve in input space but harder in F, because we cannot explicitly compute the mean of the mapped observations in F. There is, however, a way to do it.

Given any ϕ and any set of observations $x_1, ..., x_m$, let us define

$$\bar{\phi} := \frac{1}{m} \sum_{i=1}^{m} \phi(x_i)$$

then, the points

$$\bar{\phi}(x_i) = \phi(x_i) - \bar{\phi} \tag{7}$$

will be centered. Thus the assumption made above now hold, and we go on to define covariance matrix and dot product matrix $\tilde{K}_{ij} = \left\langle \bar{\phi}(x_i), \bar{\phi}(x_j) \right\rangle$ in F. We arrive at our already familiar eigenvalue problem

$$m\tilde{\lambda}\tilde{\alpha} = \tilde{K}\tilde{\alpha}, \tag{8}$$

with $\tilde{\alpha}$ being the expansion coefficients of an eigenvector (in F) in terms of the centered points (7)

$$\tilde{V} = \sum_{i=1}^{m} \tilde{\alpha}_i \bar{\phi}(x_i). \tag{9}$$

Because we do not have the centered data (7), we cannot compute $\tilde{K}$ explicitly, however we can express it in terms of its noncentered counterpart K. In the following, we shall use $K_{ij} = \left\langle \phi(x_i), \phi(x_j) \right\rangle$. To compute $\tilde{K}_{ij} = \left\langle \bar{\phi}(x_i), \bar{\phi}(x_j) \right\rangle$, we have:

$$\tilde{K}_{ij} = \left\langle \phi(x_i) - \bar{\phi}, \phi(x_j) - \bar{\phi} \right\rangle$$

$$= K_{ij} - \frac{1}{m} \sum_{t=1}^{m} K_{it} - \frac{1}{m} \sum_{s=1}^{m} K_{sj} + \frac{1}{m^2} \sum_{s,t=1}^{m} K_{st}.$$

Using the vector $\mathbf{1}_m = (1, ..., 1)^\mathsf{T}$, we get the more compact expression

$$\tilde{K} = K - \frac{1}{m} K \mathbf{1}_m \mathbf{1}_m^\mathsf{T} - \frac{1}{m} \mathbf{1}_m \mathbf{1}_m^T K + \frac{1}{m^2} (\mathbf{1}_m^\mathsf{T} K \mathbf{1}_m) \mathbf{1}_m \mathbf{1}_m^\mathsf{T}.$$

We thus can compute $\tilde{K}$ from K and solve the eigenvalue problem (8). As in equation (5), the solution $\tilde{\alpha}^k$, $k = 1, ..., r$, are normalized by normalizing the corresponding vector $\tilde{V}^k$ in F, which translates into $\tilde{\lambda}_k \left\langle \tilde{\alpha}^k, \tilde{\alpha}^k \right\rangle = 1$.

Consider a test point y. To find its coordinates we compute projections of centered ϕ-images of y onto the eigenvectors of the covariance matrix of the centered points,

$$\left\langle \bar{\phi}(y), \tilde{V}^k \right\rangle = \left\langle \phi(y) - \bar{\phi}, \tilde{V}^k \right\rangle = \sum_{i=1}^{m} \tilde{\alpha}_i^k \left\langle \phi(y) - \bar{\phi}, \phi(x_i) - \bar{\phi} \right\rangle$$

$$= \sum_{i=1}^{m} \tilde{\alpha}_i^k \left(\left\langle \phi(y), \phi(x_i) \right\rangle - \left\langle \bar{\phi}, \phi(x_i) \right\rangle - \left\langle \phi(y), \bar{\phi} \right\rangle + \left\langle \bar{\phi}, \bar{\phi} \right\rangle \right)$$

$$= \sum_{i=1}^{m} \tilde{\alpha}_i^k \left\{ K(y, x_i)) - \frac{1}{m} \sum_{s=1}^{m} K(x_s, x_i)) - \frac{1}{m} \sum_{s=1}^{m} K(y, x_s) + \frac{1}{m^2} \sum_{s,t=1}^{m} K(x_s, x_t) \right\}.$$

Introducing the vector

$$\mathbf{Z} = \Big(K(\mathbf{y}, \mathbf{x}_i)\Big)_{m \times 1}. \tag{10}$$

Then,

$$
\begin{aligned}
\Big(\langle \tilde{\phi}(\mathbf{y}), \tilde{\mathbf{V}}^k \rangle\Big)_{1 \times r} &= \mathbf{Z}^\mathsf{T} \tilde{\mathbf{V}} - \frac{1}{m} \mathbf{1}_m^\mathsf{T} K \tilde{\mathbf{V}} - \frac{1}{m}(\mathbf{Z}^\mathsf{T} \mathbf{1}_m) \mathbf{1}_m^\mathsf{T} \tilde{\mathbf{V}} + \frac{1}{m^2}(\mathbf{1}_m^\mathsf{T} K \mathbf{1}_m) \mathbf{1}_m^\mathsf{T} \tilde{\mathbf{V}} \\
&= \mathbf{Z}^\mathsf{T} \Big(\mathbf{I}_m - \frac{1}{m} \mathbf{1}_m \mathbf{1}_m^\mathsf{T}\Big) \tilde{\mathbf{V}} - \frac{1}{m} \mathbf{1}_m^\mathsf{T} K \Big(\mathbf{I}_m - \frac{1}{m} \mathbf{1}_m \mathbf{1}_m^\mathsf{T}\Big) \tilde{\mathbf{V}} \\
&= \Big(\mathbf{Z}^\mathsf{T} - \frac{1}{m} \mathbf{1}_m^\mathsf{T} K\Big) \Big(\mathbf{I}_m - \frac{1}{m} \mathbf{1}_m \mathbf{1}_m^\mathsf{T}\Big) \tilde{\mathbf{V}},
\end{aligned} \tag{11}
$$

where $\tilde{\mathbf{V}}$ is a $m \times r$ matrix whose columns are the eigenvectors $\tilde{\mathbf{V}}^1, ..., \tilde{\mathbf{V}}^r$.

Notice that the KPCA uses only implicitly the input variables since the algorithm formulates the reduction of the dimension in the feature space through the kernel function evaluation. Thus KPCA is usefulness for nonlinear feature extraction by reducing the dimension but not to explain the selected features by means the input variables.

3. Adding input variable information into Kernel PCA

In order to get interpretability we add supplementary information into KPCA representation. We have developed a procedure to project any given input variable onto the subspace spanned by the eigenvectors (9).

We can consider that our observations are realizations of the random vector $X = (X_1, ..., X_n)$. Then to represent the prominence of the input variable X_k in the KPCA. We take a set of points of the form $\mathbf{y} = \mathbf{a} + s\mathbf{e}_k \in \mathbb{R}^n$ where $\mathbf{e}_k = (0, ..., 1, ..., 0) \in \mathbb{R}^n$, $s \in \mathbb{R}$, where k-th component is equal 1 and otherwise are 0. Then, we can compute the projections of the image of these points $\phi(\mathbf{y})$ onto the subspace spanned by the eigenvectors (9).

Taking into account equation (11) the induced curve in the eigenspace expressed in matrix form is given by the row vector:

$$\sigma(s)_{1 \times r} = \Big(\mathbf{Z}_s^\mathsf{T} - \frac{1}{m} \mathbf{1}_m^\mathsf{T} K\Big) \Big(\mathbf{I}_m - \frac{1}{m} \mathbf{1}_m \mathbf{1}_m^\mathsf{T}\Big) \tilde{\mathbf{V}}, \tag{12}$$

where $\mathbf{Z}_s$ is of the form (10).

In addition we can represent directions of maximum variation of $\sigma(s)$ associated with the variable X_k by projecting the tangent vector at $s = 0$. In matrix form, we have

$$\frac{d\sigma}{ds}\Big|_{s=0} = \frac{d\mathbf{Z}_s^\mathsf{T}}{ds}\Big|_{s=0} \Big(\mathbf{I}_m - \frac{1}{m} \mathbf{1}_m \mathbf{1}_m^\mathsf{T}\Big) \tilde{\mathbf{V}} \tag{13}$$

with

$$\frac{d\mathbf{Z}_s^\mathsf{T}}{ds}\Big|_{s=0} = \Big(\frac{d\mathbf{Z}_s^1}{ds}\Big|_{s=0}, ..., \frac{d\mathbf{Z}_s^m}{ds}\Big|_{s=0}\Big)^\mathsf{T}$$

and, with

$$\frac{dZ_s^i}{ds}\bigg|_{s=0} = \frac{dK(\mathbf{y}, \mathbf{x}_i)}{ds}\bigg|_{s=0}$$

$$= \left(\sum_{t=1}^{m} \frac{\partial K(\mathbf{y}, \mathbf{x}_i)}{\partial y_t} \frac{dy_t}{ds}\right)\bigg|_{s=0}$$

$$= \sum_{t=1}^{m} \frac{\partial K(\mathbf{y}, \mathbf{x}_i)}{\partial y_t}\bigg|_{\mathbf{y}=\mathbf{a}} \delta_t^k = \frac{\partial K(\mathbf{y}, \mathbf{x}_i)}{\partial y_k}\bigg|_{\mathbf{y}=\mathbf{a}}$$

where δ_t^k denotes the delta of Kronecker. In particular, let us consider the radial basis kernel: $k(\mathbf{x}, \mathbf{z}) = \exp(-c \|\mathbf{x} - \mathbf{z}\|^2)$ with $c > 0$ a free parameter. Using above notation, we have

$$K(\mathbf{y}, \mathbf{x}_i) = \exp(-c \|\mathbf{y} - \mathbf{x}_i\|^2) = \exp\left(-c \sum_{t=1}^{n} (y_i - x_{it})^2\right)$$

When we consider the set of points of the form $\mathbf{y} = \mathbf{a} + s\mathbf{e}_k \in \mathbb{R}^n$,

$$\frac{dZ_s^i}{ds}\bigg|_{s=0} = \frac{\partial K(\mathbf{y}, \mathbf{x}_i)}{\partial y_k}\bigg|_{\mathbf{y}=\mathbf{a}}$$

$$= -2cK(\mathbf{a}, \mathbf{x}_i)(a_k - x_{ik})$$

In addition, if $\mathbf{a} = \mathbf{x}_\beta$ (a training point) then

$$\frac{dZ_s^i}{ds}\bigg|_{s=0} = -2cK(\mathbf{x}_\beta, \mathbf{x}_i)(x_{\beta k} - x_{ik})$$

Thus, by applying equation (12) we can locally represent any given input variable in the KPCA plot. Moreover, by using equation (13) we can represent the tangent vector associated with any given input variable at each sample point. Therefore, we can plot a vector field over the KPCA that points to the growing directions of the given variable.

We used the radial basis kernel in our implementation however the procedure we have introduced is also applicable to any other smooth kernel, for instance the Linear kernel which supplies standard PCA analysis.

4. Validation

In this section we illustrate our procedure with data from the leukemia data set of Golub et al. (1999) and the lymphoma data set Alizadeh et al. (2000).

In these examples our aim is to validate our procedure for adding input variables information into KPCA representation. We follow the following steps. First, in each data set, we build a list of genes that are differentially expressed. This selection is based in accordance with previous studies such as (Golub et al. (1999), Pittelkow & Wilson (2003), Reverter et al. (2010)). In addition we compute the expression profile of each gene selected, this profile confirm the evidence of differential expression.

Second, we compute the curves through each sample point associated with each gene in the list. These curves are given by the ϕ-image of points of the form:

$$\mathbf{y}(s) = \mathbf{x}_i + s\mathbf{e}_k$$

where $\mathbf{x}_i$ is the $1 \times n$ expression vector of the i-th sample, $i = 1, ..., m$, k denotes the index in the expression matrix of the gene selected to be represented, $\mathbf{e}_k = (0, ..., 1, ..., 0)$ is a $1 \times n$ vector with zeros except in the k-th. These curves describe locally the change of the sample x_i induced by the change of the gene expression.

Third, we project the tangent vector of each curve at $s = 0$, that is, at the sample points $\mathbf{x}_i$, $i = 1, ..., m$, onto the KPCA subspace spanned by the eigenvectors (9). This representation capture the direction of maximum variation induced in the samples when the expression of gene increases.

By simultaneously displaying both the samples and the gene information on the same plot it is possible both to visually detect genes which have similar profiles and to interpret this pattern by reference to the sample groups.

4.1 Leukemia data sets

The leukemia data set is composed of 3051 gene expressions in three classes of leukemia: 19 cases of B-cell acute lymphoblastic leukemia (ALL), 8 cases of T-cell ALL and 11 cases of acute myeloid leukemia (AML). Gene expression levels were measured using Affymetrix high-density oligonucleotide arrays.

The data were preprocessed according to the protocol described in Dudoit et al. (2002). In addition, we complete the preprocessing of the gene expression data with a microarray standardization and gene centring.

In this example we perform the KPCA , as detailed in the previous section, we compute the kernel matrix with using the radial basis kernel with $c = 0.01$, this value is set heuristically. The resulting plot is given in Figure 1. It shows the projection onto the two leading kernel principal components of microarrays. In this figure we can see that KPCA detect the group structure in reduced dimension. AML, T-cell ALL and B-cell ALL are fully separated by KPCA.

To validate our procedure we select a list of genes differentially expressed proposed by (Golub et al. (1999), Pittelkow & Wilson (2003), Reverter et al. (2010)) and a list of genes that are not differentially expressed. In particular, in Figures 2, 3, 4 and 5 we show the results in the case of genes: X76223_s_at, X82240_rna1_at, Y00787_s_at and D50857_at, respectively. The three first genes belong to the list of genes differentially expressed and the last gene is not differentially expressed.

Figure 2 (top) shows the tangent vectors associated with X76223_s_at gene, attached at each sample point. This vector field reveals upper expression towards T-cell cluster as is expected from references above mentioned. This gene is well represented by the second principal component. The length of the arrows indicate the strength of the gene on the sample position despite the dimension reduction. Figure 2 (bottom) shows the expression profile of

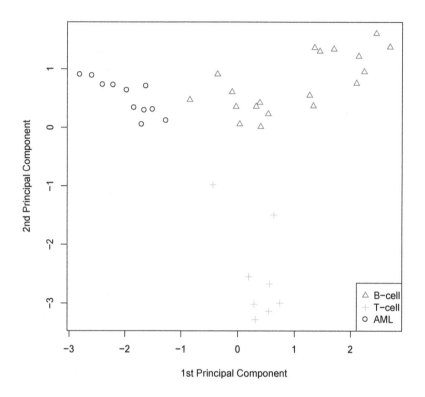

Fig. 1. Kernel PCA of Leukemia dataset.

X76223_s_at gene. We can observe that X76223_s_at gene is up regulated in T-cell class. This profile is agree with our procedure because the direction in which the expression of the X76223_s_at gene increases points to the T-cell cluster.

Figure 3 (top) shows the tangent vectors associated with X82240_rna1_at gene attached at each sample point. This vector field reveals upper expression towards B-cell cluster as is expected from references above mentioned. Figure 3 (bottom) shows the expression profile of X82240_rna1_at gene. We can observe that X82240_rna1_at gene is up regulated in B-cell class. This profile is agree with our procedure because the direction in which the expression of the X82240_rna1_at gene increases points to the B-cell cluster.

Figure 4 (top) shows the tangent vectors associated with Y00787_s_at gene attached at each sample point. This vector field reveals upper expression towards AML cluster as is expected from references above mentioned. Figure 4 (bottom) shows the expression profile of Y00787_s_at gene. We can observe that Y00787_s_at gene is up regulated in AML class. This profile is agree with our procedure because the direction in which the expression of the Y00787_s_at gene increases points to the AML cluster.

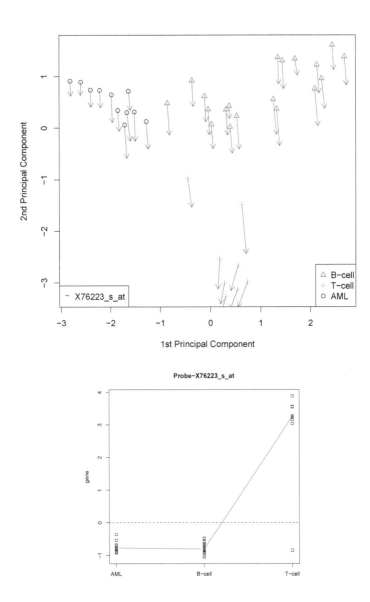

Fig. 2. (Top) Kernel PCA of Leukemia dataset and tangent vectors associated with X76223-s-at gene at each sample point. Vector field reveals upper expression towards T-cell cluster. (Bottom) Expression profile of X76223-s-at gene confirms KPCA plot enriched with tangent vectors representation.

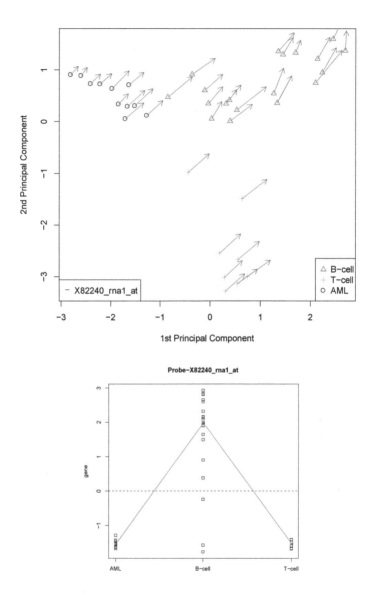

Fig. 3. (Top) Kernel PCA of Leukemia dataset and tangent vectors associated with X82240-rna1-at gene at each sample point. Vector field reveals upper expression towards B-cell cluster. (Bottom) Expression profile of X82240-rna1-at gene confirms KPCA plot enriched with tangent vectors representation.

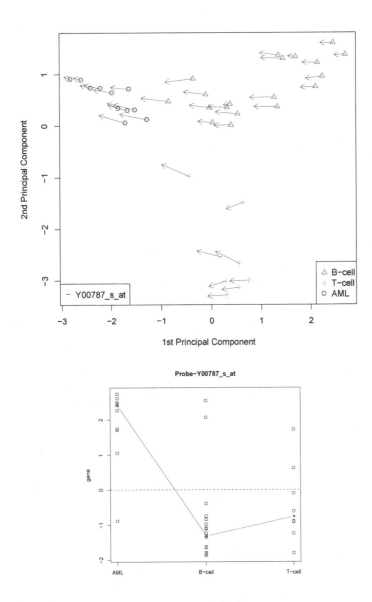

Fig. 4. (Top) Kernel PCA of Leukemia dataset and tangent vectors associated with Y00787-sat gene at each sample point. Vector field reveals upper expression towards AML cluster. (Bottom) Expression profile of Y00787-sat gene confirms KPCA plot enriched with tangent vectors representation..

Figure 5 (top) shows the tangent vectors associated with D50857_at gene attached at each sample point. This vector field shows no preferred direction to any of the three cell groups. The arrows are of short length and variable direction in comparison with other genes showed in previous Figures. Figure 5 (bottom) shows a flat expression profile of D50857_at gene. This profile is agree with our procedure because any direction of expression of the D50857_at gene is highlighted.

4.2 Lymphoma data sets

The lymphoma data set comes from a study of gene expression of three prevalent lymphoid malignancies: B-cell chronic lymphocytic leukemia (B-CLL), follicular lymphoma (FL) and diffuse large B-cell lymphoma (DLCL). Among 96 samples we took 62 samples 4026 genes in three classes: 11 cases of B-CLL, 9 cases of FL and 42 cases of DLCL. Gene expression levels were measured using 2-channel cDNA microarrays.

After preprocessing, all gene expression profiles were base 10 log-transformed and, in order to prevent single arrays from dominating the analysis, standardized to zero mean and unit variance. Finally, we complete the preprocessing of the gene expression data with gene centring.

In this example we perform the KPCA , as detailed in the previous section, we compute the kernel matrix with using the radial basis kernel with $c = 0.01$, this value is set heuristically. The resulting plot is given in Figure 6. It shows the projection onto the two leading kernel principal components of microarrays. In this figure we can see that KPCA detect the group structure in reduced dimension. DLCL, FL and B-CLL are fully separated by KPCA.

To validate our procedure we select a list of genes differentially expressed proposed by (Reverter et al. (2010)) and a list of genes that are not differentially expressed. In particular, in Figures 7, 8, 9 and 10 we show the results in the case of genes: 139009, 1319066, 1352822 and 1338456, respectively. The three first genes belong to the list of genes differentially expressed and the last gene is not differentially expressed.

Figure 7 (top) shows the tangent vectors associated with 139009 gene attached at each sample point. This vector field reveals upper expression towards DLCL cluster as is expected from references above mentioned. This gene is mainly represented by the first principal component. The length of the arrows indicate the influence strength of the gene on the sample position despite the dimension reduction. Figure 7 (bottom) shows the expression profile of 139009 gene. We can observe that 139009 gene is up regulated in DLCL cluster. This profile is agree with our procedure because the direction in which the expression of the 139009 gene increases points to the DLCL cluster.

Figure 8 (top) shows the tangent vectors associated with 1319066 gene attached at each sample point. This vector field reveals upper expression towards FL cluster as is expected from references above mentioned. This gene is mainly represented by the second principal component. Figure 8 (bottom) shows the expression profile of 1319066 gene. We can observe that 1319066 gene is up regulated in FL class. This profile is agree with our procedure because the direction in which the expression of the 1319066 gene points to the FL cluster.

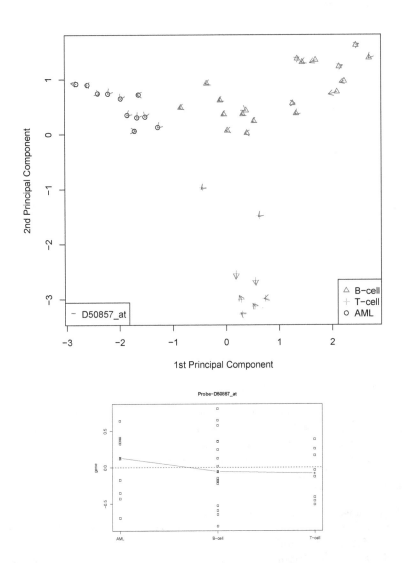

Fig. 5. (Top) Kernel PCA of Leukemia dataset and tangent vectors associated with D50857-at gene at each sample point. Vector field shows no preferred direction. (Bottom) Flat Expression profile of D50857-at gene confirms KPCA plot enriched with tangent vectors representation.

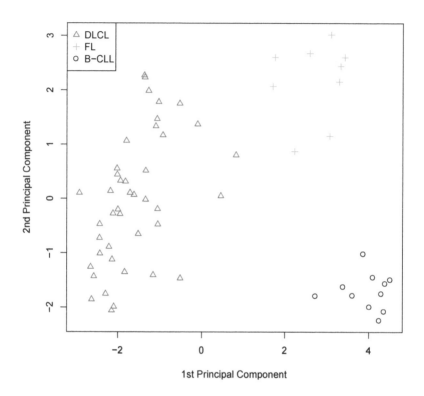

Fig. 6. Kernel PCA of Lymphoma dataset.

Figure 9 (top) shows the tangent vectors associated with 1352822 gene attached at each sample point. This vector field reveals upper expression towards B-CLL as is expected from references above mentioned. Figure 9 (bottom) shows the expression profile of 1352822 gene. We can observe that 1352822 gene is up regulated in B-CLL class. This profile is agree with our procedure because the direction in which the expression of the 1352822 gene increases points to the B-CLL cluster.

Figure 10 (top) shows the tangent vectors associated with 1338456 gene attached at each sample point. This vector field shows no preferred direction to any of the three cell groups. The arrows are of short length and variable direction in comparison with other genes showed in previous Figures. Figure 10 (bottom) shows a flat expression profile of 1338456 gene. This profile is agree with our procedure because any direction of expression of the 1338456 gene is highlighted.

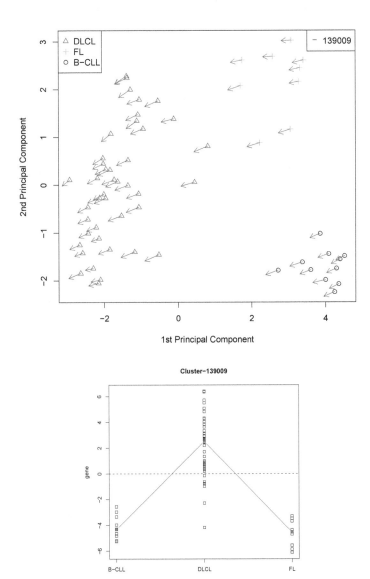

Fig. 7. (Top) Kernel PCA of Leukemia dataset and tangent vectors associated with 139009 gene at each sample point. Vector field reveals upper expression towards DLCL cluster. (Bottom) Expression profile of 139009 gene confirms KPCA plot enriched with tangent vectors representation.

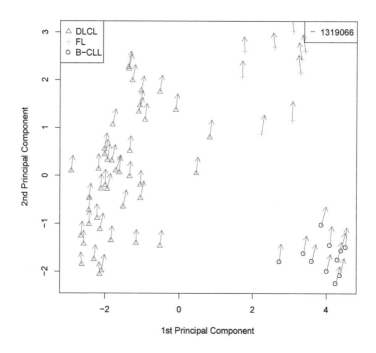

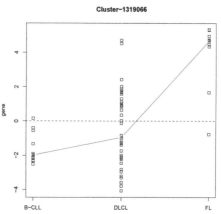

Fig. 8. (Top) Kernel PCA of Leukemia dataset and tangent vectors associated with 1319066 gene at each sample point. Vector field reveals upper expression towards FL cluster. (Bottom) Expression profile of 1319066 gene confirms KPCA plot enriched with tangent vectors representation.

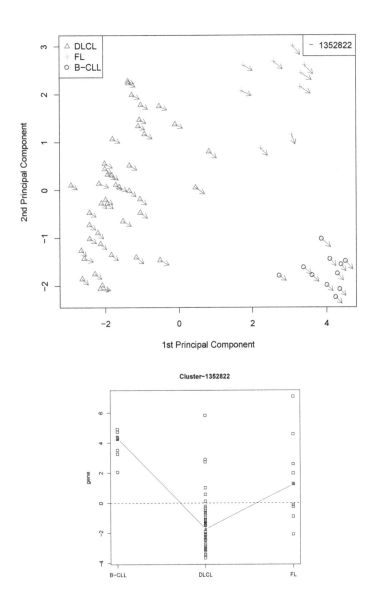

Fig. 9. (Top) Kernel PCA of Leukemia dataset and tangent vectors associated with 1352822 gene at each sample point. Vector field reveals upper expression towards B-CLL cluster. (Bottom) Expression profile of 1352822 gene confirms KPCA plot enriched with tangent vectors representation.

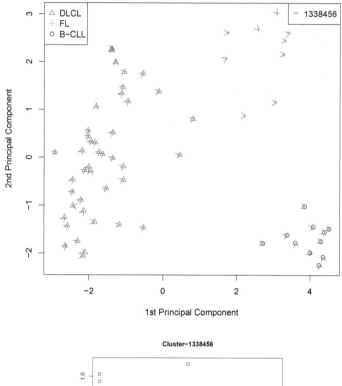

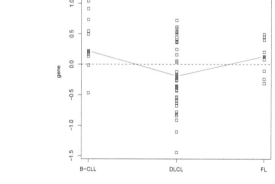

Fig. 10. (Top) Kernel PCA of Leukemia dataset and tangent vectors associated with 1338456 gene at each sample point. Vector field shows no preferred direction. (Bottom) Flat expression profile of 1338456 gene confirms KPCA plot enriched with tangent vectors representation.

5. Conclusion

In this paper we propose an exploratory method based on Kernel PCA for elucidating relationships between samples (microarrays) and variables (genes). Our approach show two main properties: extraction of nonlinear features together with the preservation of the input variables (genes) in the output display. The method described here is easy to implement and facilitates the identification of genes which have a similar or reversed profiles. Our results indicate that enrich the KPCA with supplementary input variable information is complementary to other tools currently used for finding gene expression profiles, with the advantage that it can capture the usual nonlinear nature of microarray data.

6. References

Alizadeh, A.A.; Eisen, M.B.; Davis, R,E.; Ma, C.; Lossos, I.S.; Rosenwald, A.; Boldrick, J.C.; Sabet, H.; Tran, T.; Yu, X.; Powell, J.I.; Yang, L.; Marti, G.E.; Moore, T.; Hudson, J.J.; Lu, L.; Lewis, D.B.; Tibshirani, R.; Sherlock, G.; Chan, WC.; Greiner, T.C.; Weisenburger, D.D.; Armitage, J.O.; Warnke, R.; Levy, R.; Wilson, W.; Grever, M.R.; Byrd, J.C; Bostein, D.; Brown, P,O. & Staudt, LM. (2000). Different type of diffuse large b-cell lymphoma identified by gene expression profiling. *Nature*, 403:503–511.

Alter, O.; Brown, P,O. & Botstein, D. (2000). Singular value decomposition for genome-wide expression data processing and modeling. *Proc. Natl. Acad. Sci. USA.* 97(18), 10101-10106.

Chambers, J.M.; Cleveland, W.S.; Kleiner, B. & Tuckey, P.A. (1983) *Graphical Methods for Data Analysis statistics/probability.* Wadsworth.

Chu, W.; Ghahramani, Z.; Falciani. F. & Wild, D: (2005) Biomarker discovery in microarrays gene expression data with Gaussian processes. *Bioinformatcis.* 21(16), 3385–3393.

Dettling, M. (2004). BagBoosting for tumor classificcation with gene expression data. *Bioinformatcis.* 20(18), 3583–3593.

Diaz-Uriarte, R. & Andres, S.A. (2006) Gene selection and classification of microarray data using random forest. *BMC Bioinformatcis.* 7:3, 1–13.

Dudoit, S.; Fridlyand, J. & Speed, T,P. (2002). Comparison of discrimination methods for the classification of tumours using gene expression data. *J. Am. Statist. Soc.* 97:77–87.

Eisen, M.B.; Spellman, P,T.; Brown, P.O. & Botstein, D. (1998). Cluster analysis and display of genome-wide expression patterns. *Proc. Natl. Acad. Sci. USA.* 95(25):14863–14868.

Fellenberg, K.; Hauser, N.C.; Brors, B.; Neutzner, A. & Hoheisel, J. (2001): Correspondence analysis applied to microarray data. *Proc. Natl. Acad. Sci. USA.* 98(19) 10781–10786.

Golub, T.R.; Slonim, D.K.; Tamayo, P.; Huard, C.; Gaasenbeek, M.; Mesirov, J.P.; Coller, H.; Loh, M.L.; Downing, J.R.; Caligiuri, M.A.; Bloomfield, C.D. & Lander, E.S. (1999). Molecular classification of cancer: Class discovery and class prediction by gene expression profiling. *Science.* 286(5439):531–537.

Park, M.; Lee, J.W.; Lee, J.B. & Song, S.H. (2008) Several biplot methods applied to gene expression data. *Journal of Statistical Planning and Inference.* 138:500–515.

Pittelkow, Y.E. & Wilson, S.R. (2003). Visualisation of Gene Expression Data - the GE-biplot, the Chip-plot and the Gene-plot. *Statistical Applications in Genetics and Molecular Biology.* Vol. 2. Issue 1. Article 6.

Reverter, F.; Vegas, E. & Sanchez, P. (2010) Mining Gene Expressions Profiles: An integrated implementation of Kernel Principal Components Analysis and Singular Value Decomposition. *Genomics, Proteomics and Bioinformatics.* 8(3):200–210.

Shawe-Taylor, J. & Cristianini, N. (2004). *Kernel Methods for Pattern Analysis.* Cambridge University Press.

Scholkopf, B.; Smola, A.J. & Muller, K.R. (1998) Nonlinear Component Analysis as a Kernel Eigenvalue Problem. *Neural Computation.* 10:1299-1319.

Scholkopf, B.; Smola, A.J. (2002). *Learning with Kernels - Support Vector Machines, Regularization, Optimization and Beyond.* Cambridge, MA. MIT Press.

Tamayo, P.; Solni, D.; Mesirov, J.; Zhu, Q.; Kitareewan, S.; Dmitrovsky, E.; Lander, E.S. & Golub, T.R. (1999). Interpreting patterns of gene expression with self-organizing maps: Methods and application to hematopoietic differentiation. *Proc. Natl. Acad. Sci.* USA. 96(6):2907–2912.

Xuehua, L. & Lan, S. (2009). Kernel based nonlinear dimensionality reduction for microarray gene expression data analysis. *Expert Systems with Applications* 36:7644-7650.

Zhao, X. & Cheung, L.W.K. (2007) Kernel-Imbedded Gaussian processes for disease classification using microarrays gene expression data. *BMC Bioinformatcis.* 8:67:1–26.

Zhenqiu, L., Dechang, C. & Halima B. (2005). Clustering gene expression data with kernel principal components. *Journal of Bioinformatics and Computational Biology.* 3(2):303–316.

3

Empirical Study: Do Fund Managers Herd to Counter Investor Sentiment?

Tsai-Ling Liao[1], Chih-Jen Huang[1] and Chieh-Yuan Wu[2]
[1]Providence University, Taichung,
[2]Department of the Treasury Taichung Bank,
Taiwan

1. Introduction

Behavior among investors often influences one another. Investors may forego their own rational analysis but instead adopt behavior that is similar to the group. Recently, several studies note that this herding phenomenon (simultaneously trade the same stocks in the same direction) even exists in the behavior of institutional investors (Choe, Kho, and Stulz, 1999; Grinblatt, Titman, and Wermers, 1995; Kyrolainen and Perttunen, 2003; Lakonishok, Shleifer, and Vishny, 1992; Walter and Weber, 2006; Wermers, 1999; Wylie, 2005). There are three theoretical foundations for explaining institutional investor herding. First, in order to maintain or build a reputation when markets are imperfectly informed, managers may ignore their private information and hide in the herd (Brandenburger and Polak, 1996; Prendergast, 1993; Scharfstein and Stein, 1990; Trueman, 1994; Zwiebel, 1995). Second, managers may infer private information from the prior actions of agents (peer-group effects), and optimally decide to act alike (Bala and Goyal, 1998; Fung and Hsieh, 1999; Pound and Shiller, 1989). Third, institutional investors may receive the correlative private information from analyzing the same indicators (Banerjee, 1992; Bikhchandani, Hirshleifer, and Welch, 1992; Block, French, and Maberly, 2000; Froot, Scharfstein, and Stein, 1992; Hirshleifer, Subrahmanyam, and Titman, 1994) or favoring securities with specific characteristics (Del Guercio, 1996; Falkenstein, 1996; Gompers and Metrick, 2001; Payne, Prather, and Bertin, 1999).

What are the other factors that may explain the behavior of institutional investors herding? Investor sentiment may be one of the significant factors that cause this behavior. For example, Lakonishok et al. (1992, p.26) state in their pioneer research that, "...they (fund managers) might herd if they all counter the same irrational moves in individual investor sentiment." Barberis and Shleifer (2003), De Long et al. (1990), Lee et al. (1991), and Shleifer (2000) also make similar statements. Accordingly, one may intuitively expect that institutional investors will herd in their sell (buy) decisions in the presence of optimistic (pessimistic) sentiment. However, the literature appears to include no research that conducts an empirical investigation on this interesting hypothesis. The purpose of this study is thus to examine whether prior investor sentiment cross-sectionally explains the level of fund manager herding.

This study focuses on the U.S. fund trading sample because herd behavior becomes increasingly important when large institutional investors dominate the market such as in the U.S. To estimate herding by fund managers, the study in this report applies the Lakonishok et al.'s (1992) measure and Wylie's (2005) trinomial-distribution approach which considers buy, hold and sell positions for a given stock in a period. The study calculates conditional herding measures for stock-months that have a higher or lower proportion of fund buyers relative to the expected proportion in each month, which are the buy-herding measure and the sell-herding measure. To estimate investor sentiment for each stock-month, this paper employs the principal component analysis as the means of extracting the composite unobserved sentiment measure, rather than just selects a single indicator to proxy sentiment (Fisher and Statman, 2000; Neal and Wheatley, 1998). Brown and Cliff (2004) indicate that this estimating procedure is able to successfully extract measures of unobserved sentiment from various indicators. According to Baker and Stein (2004), Brown and Cliff (2004) and Baker and Wurgler (2006), the present study examines ten market weather vanes that can be categorized into three groups: individual stock sentiment indicator (individual stock return and individual stock trading volume), overall market sentiment indicator (S&P500 index return, S&P500 index trading volume, S&P500 index option Put/Call ratio, the number of IPOs, the average first-day return on IPOs, NYSE share turnover, and Russell 2000 index return), and fund sentiment indicator (net purchases of mutual funds).

After excluding observations with incomplete data to estimate herding, the final sample contains 770 U.S. mutual funds' holding records on 527 stocks from January 2001 to December 2005. Incorporating both bear (roughly 2001-2002) and bull (roughly 2003-2005) market periods under examination helps control for the effect of market conditions on the empirical results. The findings indicate a significantly positive association between the sentiment measure and subsequent sell-herding, after controlling the fund performance deviation, the capitalization of the stock, the number of funds trading the stock, the net mutual fund redemption of the stock, and the market-to-book ratio of the stock. However, the evidence shows no significant correlation between the composite sentiment measure and subsequent buy-herding. These findings suggest that institutional investors herd on the selling side when they observe high level of investor optimism, consistent with the intuition that rational institutional investors tend to counteract the optimistic sentiment of the investors (Wermers, 1999).

In sum, this paper contributes to the literature in three ways. First, by extending prior research that examines the impacts of fund herding on subsequent stock returns (Grinblatt et al., 1995; Klemkosky, 1977; Wermers, 1999), and the short-run predictability of sentiment in stock returns (Baker and Wurgler, 2006; Brown and Cliff, 2004; Fisher and Statman, 2000), this article substantiates the sentiment countering hypothesis and thereby contributes to the research realm of negative-feedback strategies, i.e. selling stocks that rise too far (Jegadeesh and Titman, 1993; Nofsinger and Sias, 1999; Scharfstein and Stein, 1990; Wermers, 1997, 1999). Second, the empirical finding indicates an explanatory power of investor sentiment for fund manager herding, providing support for the informational cascade theory where fund managers herd as a result of analyzing the sentiment indicator in this case. In other words, fund managers herd sell because, at least in part, they observe and try to counteract the optimistic sentiment. Finally, the significant association between sentiment and fund manager herding suggests potential directions for further research; specifically, the effect of fund herding behavior of countering optimistic market sentiment on fund performance. Also, researchers may employ the Structural Equation approach to examine the mediating role of institutional investor herding in the causal relationship between investor sentiment and stock returns. Such investigations will help corroborate the beneficial role of fund herding in speeding the price adjustment, as Wermers (1999) suggests, in the presence of mispricing resulting from investor sentiment.

This study proceeds in four sections. The next section discusses the estimation processes of fund herding and investor sentiment, the test methods and the variables. Section 3 describes the sample and data collection. Section 4 reports the empirical results. Finally, Section 5 summarizes the main conclusions.

2. Methods

2.1 Trinomial-distribution herding measure

To measure fund herding in stock i in month t (HMi,t), the present report follows the Lakonishok et al. (1992) method:

$$HMi,t = |Pi,t - E(Pi,t)| - E|Pi,t - E(Pi,t)|, \tag{1}$$

where,

$$Pi,t = B_{i,t} / (B_{i,t} + S_{i,t}) \tag{2}$$

$$E(P_{i,t}) = \frac{\sum_{i=1}^{n} B_{i,t}}{\sum_{i=1}^{n}(B_{i,t} + S_{i,t})} \tag{3}$$

$B_{i,t}$ ($S_{i,t}$) is the number of fund managers who buy (sell) stock i in month t. Pi,t is the proportion of mutual funds buying stock i in month t relative to the total number of funds trading stock i in month t. $E(Pi,t)$ is the sample estimate of Pi,t, the expected proportion of buyers for stock i in month t. $E|Pi,t - E(Pi,t)|$ is an adjustment which controls for random variation around the expected proportion of buyers under the null hypothesis of no herding. $|Pi,t - E(Pi,t)|$ will be large if the trading of managers polarizes in the direction of either buying or selling. Averaging HM over all stock-months of interest can measure the extent to which any fund herds in a given stock-month.

To examine mutual funds' reaction to signals conveying upward or downward consensus, the study calculates conditional herding measures for stock-months that have a higher or lower proportion of buyers than the average stock in the same month as follows:

$$BHM = HMi,t|Pi,t > E[Pi,t] \tag{4}$$

$$SHM = HMi,t|Pi,t < E[Pi,t] \tag{5}$$

This approach is useful in analyzing fund herding into stocks separately from herding out of stocks. With Equation (4) as an example, the buy-herding measure (*BHM*) is equal to *HM* conditional on $P_{i,t} > E[P_{i,t}]$. If mutual funds tend to herd in their selling of stocks more frequently than in their buying of stocks, then the sell-herding measure (SHM) will be larger than the buy-herding measure (BHM).

Previous studies (e.g., Choe et al., 1999; Wermers, 1999) focus mostly on the binomial distribution of herding measure and consider only the buy and sell positions. Wylie (2005) claims that the accuracy of the Lakonishok et al.'s (1992) measure may be in question because the analysis rests on the assumption that all fund managers can short sell all stocks. However, the U.S. law technically neither forbids nor allows mutual funds to undertake short sales. As Wylie (2005) notes, Section 12(a) of the Investment Companies Act 1940 prohibits short sales by registered mutual funds in contravention of SEC rules, while the SEC issues no rules under Section 12(a). As a result, few U.S. mutual funds undertake any short selling. Wylie thereby argues that essentially herding could arise because of this invalid assumption. To avoid obtaining systematic herding estimate, the following tests apply Wylie's (2005) trinomial-distribution approach to measure fund manager herding, including buy, hold and sell positions.

2.2 Investor sentiment measure: Factor analysis approach

This study next conducts the principal component analysis to extract indicators explaining investor sentiment. We use an orthogonal transformation to convert 10 possibly correlated indicators of investor sentiment, which are taken from the literature, into a set of values of uncorrelated indicators. We then base on the factor analysis to calculate the corresponding factor score for every sentiment proxy. The composite sentiment measure is the sum of multiplying the transformed factor score (weighting) by each indicator's value. Table 1 summarizes the definition and the corresponding weighting for each proxy. These ten sentiment indicators can be classified into three groups, including individual stock sentiment indicator, overall market sentiment indicator, and fund sentiment indicator.

2.2.1 Individual stock sentiment indicator

a. Individual stock return (ISR): De Long et al. (1990) and De Bondt (1993) find that investors predict the trend of stock returns according to past stock returns, so that the continuity of forecastable stock returns arises. For example, when the return continues for a period of time, the bullish psychology will pervade the investors. Notwithstanding the bull quotation occurs for a period of time and investors believe that the overall market would revise the quotation in the short term, the investors' bullish psychology remains.

b. Individual stock trading volume (ISV): Baker and Stein (2004) suggest that the market liquidity, like spreads, depth, trading volume and turnover, carrying information about the market can serve as a sentiment indicator. The trading volume is used as one of the sentiment proxy accordingly.

2.2.2 Overall market sentiment indicator

a. S&P500 index return ($SP500R$) and Russell 2000 index return ($RU2000R$): The reason for incorporating these two proxies is the same as the ISR.

b. S&P500 index trading volume ($SP500V$): The reason for this proxy is the same as the ISV.

c. S&P500 index option put to call ratio (*SP500PC*): Brown and Cliff (2004) note that a higher (lower) put to call ratio indicates the pessimistic (optimistic) atmosphere getting stronger. The S&P500 index option put to call ratio is included as a proxy of the market sentiment.

d. Initial public offering first-day returns (*IPOR*) and the number of offerings (*IPON*): Due to the information asymmetries between IPO managers and outside investors, IPO activity tends to occur during bullish periods. Following Brown and Cliff (2004) and Baker and Wurgler (2006), the report includes both the IPO first-day returns and the number of offerings.

e. NYSE share turnover (*NYSET*): NYSE share turnover is the ratio of reported share volume to average shares listed from the *NYSE Fact Book*. In a market with short-sales constraints, Baker and Stein (2004) indicate that irrational investors will only be active in the market when they are optimistic and high liquidity is, as a result, a symptom of overvaluation.

2.2.3 Fund sentiment indicator

Net purchases of mutual funds (*NP*): Brown and Cliff (2004) indicate that net purchases by mutual funds and foreign investors positively reflect the extent to which they are optimistic about the individual stock. *NP* is the difference between dollar purchases and dollar sales by fund managers in a given stock.

	Indicator	Definition	Weighting
Individual stock sentiment	Individual stock return (ISR)	$\text{Log}(P_{i,t}) - \text{Log}(P_{i,t-1})$	0.097
	Individual stock trading volume (ISV)	$\text{Log}(V_{i,t}) - \text{Log}(V_{i,t-1})$	0.073
Market sentiment	S&P500 index return (SP500R)	$\text{Log}(SP_t) - \text{Log}(SP_{t-1})$	0.055
	Russell 2000 index return (RU2000R)	$\text{Log}(Ru_t) - \text{Log}(Ru_{t-1})$	0.089
	S&P500 index trading volume (SP500V)	$\text{Log}(SPV_t) - \text{Log}(SPV_{t-1})$	0.093
	S&P500 index option Put/Call ratio variation (SP500PC)	$(\text{Put}_t/\text{Call}_t) - (\text{Put}_{t-1}/\text{Call}_{t-1})$	0.105
	Initial public offering first-day return (IPOR)	See Jay Ritter website	0.152
	Number of offerings (IPON)	See Jay Ritter website	0.146
	NYSE share turnover (NYSET)	Volume/shares listed	0.116
Mutual fund managers sentiment	Net purchases of mutual funds (NP)	$B_t - S_t$	0.074

Table 1. Investor Sentiment Indicators

Note. $P_{i,t}$ is the closing price at the end of month t for stock i. $V_{i,t}$ is the trading volume of stock i in month t. SP_t is the closing price at the end of month t for S&P500 index. Ru_t is the closing price at the end of month t for Russell 2000 index. SPV_t is the S&P500 trading volume in month t. Call_t (Put_t) is the open interest of S&P500 call (put) option in month t. NP is the difference between dollar purchases and dollar sales by funds in a given stock, where B_t (S_t) is the dollar purchases (sales) by funds in stock i in month t. The weighting for each sentiment indicator is the transformed factor score by principal component analysis.

2.3 The effect of sentiment on fund herding

To gauge the effect of investor sentiment on fund herding, the report estimates the following regression equations:

$$HM_{i,t}=\alpha_0+\alpha_1*S_{i,t-1}+\alpha_2*RSTD_{t-1}+\alpha_3*CAP_{i,t}+\alpha_4*TNUMBER_{i,t}+\alpha_5*PR_t+\alpha_6*PB_{i,t-1}+\varepsilon_{i,t} \qquad (6)$$

$$BHM_{i,t}=\alpha_0+\alpha_1*S_{i,t-1}+\alpha_2*RSTD_{t-1}+\alpha_3*CAP_{i,t}+\alpha_4*TNUMBER_{i,t}+\alpha_5*PR_t+\alpha_6*PB_{i,t-1}+\varepsilon_{i,t} \qquad (7)$$

$$SHM_{i,t}=\alpha_0+\alpha_1*S_{i,t-1}+\alpha_2*RSTD_{t-1}+\alpha_3*CAP_{i,t}+\alpha_4*TNUMBER_{i,t}+\alpha_5*PR_t+\alpha_6*PB_{i,t-1}+\varepsilon_{i,t} \qquad (8)$$

Here, *HM* is the aggregate fund herding measure, *BHM* is the buy-herding measure and *SHM* is the sell-herding measure. The main dependent variable, $S_{i,t-1}$, is the composite sentiment index of stock i in month $t-1$. To support the sentiment countering argument, α_1 is expected to be positive.

As for control variables, $RSTD_{t-1}$ is the standard deviation of fund returns in month $t-1$. Bikhchandani et al. (1992) state that herding may result from some fund managers following other leader funds in their trades. A larger return deviation among funds in previous period indicates managers sharing dissimilar information and implies a higher reputation risk for underperformed managers, giving mutual funds a greater incentive to herd. Therefore, α_2 is expected to be positive. Lakonishok et al. (1992), Wermers (1999) and Choe et al. (1999) establish that managers herd more on small stocks. $CAP_{i,t}$ is defined as the capitalization of stock i in month t, and the coefficient (α_3) is expected to be negative. Wermers (1999) and Wylie (2005) indicate a positive relation between the level of herding in a given stock and the number of funds trading that stock ($TNUMBER_{i,t}$). The coefficient of $TNUMBER$ (α_4) is thus expected to be positive.

Wermers (1999) argues that fund may herd in response to sudden increases in cash inflow (mutual fund purchases) or cash outflows (mutual fund redemptions). The disproportional purchase (redemption) waves initiated by fund investors may force managers to increase (decrease) their stock holding simultaneously, and hence result in a buy (sell) herding. The net mutual fund redemption (PR_t) is the ratio of mutual fund purchases minus mutual fund redemptions in month t to fund assets. The coefficient (α_5) is expected to be positive (negative) when *BHM* (*SHM*) is the dependent variable of the regression equation. The ratio of market value to book equity at the beginning of the trading quarter for stock i ($PB_{i,t-1}$) is used as a proxy for growth opportunity. The coefficient (α_6) is expected to be positive as Wermers (1999) shows that higher levels of herding by growth-oriented funds are consistent with growth fund possessing less precise information about the true value of their stockholdings (mainly growth stocks). Finally, an industry dummy (equal to one for stocks of high-tech firms, zero otherwise) and four yearly dummies (2001 as the base year) are included in Equations (6) through (8) to control the possible industry and macroeconomic effects on fund herding behavior.

3. Sample and data collection

The sample contains 770 U.S. mutual funds' trading records on 527 stocks from 2003 to 2007, after excluding observations with incomplete data to estimate fund herding. Taking each monthly change in a stock as a separate trading, the total stock-month observations amount to 31,093, including 17,095 buy-herding and 13,998 sell-herding observations. Instead of using quarterly fund trading data, this study analyzes the fund herding behavior with monthly data to reduce the impact of intra-period round-trip transactions on the results.

Monthly data on portfolio holdings of the U.S. mutual funds are collected from the CRSP Mutual Fund Database. These data include periodic share holdings of equities for each fund at the end of each month. Individual stock price, trading volume and capitalization are obtained from the CRSP as well. S&P500 index return, S&P500 trading volume, S&P500 index option call/put open interest, NYSE share turnover and Russell 2000 index return are compiled from the Datastream. The number of IPOs and the average first-day returns of IPOs are available at the Jay Ritter website (http://bear.cba.ufl.edu/ritter).

4. Empirical results

4.1 Descriptive statistics and correlations between selected variables

Table 2 provides the descriptive statistics of the three herding measures, ten sentiment-related indicators and selected control variables in the regression analyses. The mean herding measures are 1.779% (HM), 2.164% (BHM) and 1.308% (SHM), respectively. The medians of the three herding measures are even smaller. These statistics indicate that fund managers seem not to herd much, in line with the findings of Lakonishok et al. (1992) and werners (1999).

Variable	Mean	Std. Dev.	Min	Q1	Median	Q3	Max
HM (%)	1.779	5.606	-16.232	-1.612	0.287	3.798	97.664
BHM (%)	2.164	5.913	-16.648	-1.577	0.548	4.335	77.654
SHM (%)	1.308	5.267	-16.233	-1.686	0.196	3.280	96.685
ISR	0.007	0.040	-0.614	-0.013	0.009	0.031	0.375
ISV	0.046	0.055	-0.085	-0.038	-0.006	0.040	1.142
SP500R	0.006	0.070	-0.015	-0.003	0.005	0.007	0.025
SP500V	0.041	0.049	-0.079	-0.041	0.018	0.058	0.101
RU2000R	0.008	0.042	-0.068	-0.022	0.021	0.044	0.089
SP500PC	-0.012	0.076	-0.147	-0.034	-0.025	0.030	0.139
IPOR	0.122	0.013	0.102	0.121	0.122	0.123	0.140
IPON	142.600	45.026	63	156	159	161	174
NYSET	0.432	0.421	0.363	0.621	0.625	0.632	0.648
NP	267,447	33,103,284	-1.079E9	-1,970,434	126,690	2,391,904	601,787,572
S_{t-1}	54.961	13.164	33.969	47.003	56.552	60.168	69.347
$RSTD_{t-1}$	0.022	0.013	0.015	0.016	0.022	0.028	0.045
CAP_t	6,550,706	21,769,725	42,230	557,247	1,417,585	4,107,318	317,133,094
$TNUMBER_t$	120.166	121.391	2	56	87	146	651
PR_t	0.004	0.004	-0.015	0.002	0.004	0.009	0.017
PB_{t-1}	7.970	213.206	-136.187	1.594	2.621	3.966	13372.830

Table 2. Descriptive Statistics of Herding Measures and Selected Variables

Note. The sample contains 770 U.S. mutual funds' holding records on 527 stocks from 2001 to 2005, including 31,093 (17,095 buy-herding and 13,998 sell-herding) observations. HM is the aggregate herding measure. BHM is the buy-herding measure. SHM is the sell-herding measure. ISR is the individual stock return. ISV is the change of the logarithm of individual stock trading volume. SP500R is the S&P500 index return. SP500V is the change of the logarithm of S&P500 index trading volume.

RU2000R is the Russell 2000 index return. SP500PC is the S&P500 index option Put/Call ratio variation. IPOR is the mean first-day return of IPOs. IPON is the number of IPOs. NYSET is the NYSE share turnover. NP is the difference between dollar purchases and dollar sales by fund managers in a given stock. S is the composite sentiment measure. RSTD is the standard deviation of fund return. CAP is the capitalization of a given stock. TNUMBER is the number of funds trading a given stock. PR is the ratio of mutual fund purchases minus mutual fund redemptions in a given month to net assets of the fund. PB is the ratio of market value to book equity at the beginning of the trading quarter.

Table 3 presents Spearman (upper-triangle) and Pearson (lower-triangle) correlation coefficients between selected variables. In the Spearman measure, the correlation between the composite sentiment measure (S_{t-1}) and the aggregate herding (HM_t) is significantly positive (0.032). Likewise, the correlation between S_{t-1} and the sell-herding (SHM_t) is also significantly positive (0.039). However, the correlation between S_{t-1} and the buy-herding (BHM_t) is not significantly different from zero (0.013). The bivariate correlation evidence shows that investor sentiment does affect fund herding, especially on the sell-side. Although some of the figures are insignificant, the correlation coefficients between control variables and the herding measures correspond mostly with the predicted signs. The results in the Pearson measure are qualitatively similar to the findings in the Spearman measure.

	Panel A: Herding measure (HM)						
	HM_t	S_{t-1}	$RSTD_{t-1}$	CAP_t	$TNUMBER_t$	PR_t	PB_{t-1}
HM_t		0.032***	0.066***	0.005	0.012	0.010	0.019
S_{t-1}	0.031**		-0.304***	0.035***	0.043***	0.160***	0.019
$RSTD_{t-1}$	0.066***	-0.414***		-0.007	-0.077***	0.092***	-0.013
CAP_t	-0.014	0.028**	-0.001		0.551***	0.030***	0.014
$TNUMBER_t$	0.026**	0.046***	-0.058***	0.433***		-0.008	0.007
PR_t	0.016	0.170***	0.134***	0.022*	-0.027**		0.006
PB_{t-1}	0.005	0.009	-0.016	-0.005	-0.014	-0.008	
	Panel B: Buy-herding measure (BHM)						
	BHM_t	S_{t-1}	$RSTD_{t-1}$	CAP_t	$TNUMBER_t$	PR_t	PB_{t-1}
BHM_t		0.013	0.032***	-0.065***	0.046***	0.034***	-0.007
S_{t-1}	-0.013		-0.285***	0.028**	0.014	0.226***	0.019
$RSTD_{t-1}$	0.027**	-0.411***		0.037***	0.019	-0.188***	0.011
CAP_t	-0.034**	0.011	0.086***		0.474***	0.110***	0.005
$TNUMBER_t$	0.073***	0.016	0.052***	0.412***		0.061***	-0.004
PR_t	0.050***	0.198***	-0.021*	0.094***	0.081***		0.008
PB_{t-1}	0.005	0.013	-0.024*	-0.005	-0.017	-0.008	
	Panel C: Sell-herding measure (SHM)						
	SHM_t	S_{t-1}	$RSTD_{t-1}$	CAP_t	$TNUMBER_t$	PR_t	PB_{t-1}
SHM_t		0.039***	0.123***	0.113***	0.121***	-0.066***	0.026*
S_{t-1}	0.026**		-0.300***	0.067***	0.075***	0.117***	0.008
$RSTD_{t-1}$	0.101***	-0.443***		-0.083***	-0.170***	0.156***	-0.022
CAP_t	0.049***	0.037***	-0.073***		0.522***	-0.078***	0.017
$TNUMBER_t$	0.039***	0.086***	-0.150***	0.455***		-0.114***	0.010
PR_t	-0.098***	0.126***	0.242***	-0.060***	-0.104***		0.007
PB_{t-1}	0.016	-0.001	0.003	-0.002	-0.007	-0.016	

Table 3. Correlation Coefficients between Regression Variables

Note. The Spearman (Pearson) correlation coefficients are reported at the upper-triangle (lower-triangle). The sample contains 770 U.S. mutual funds' holding records on 527 stocks from 2001 to 2005, including 31,093 (17,095 buy-herding and 13,998 sell-herding) stock-month observations. HM is the aggregate herding measure. BHM is the buy-herding measure. SHM is the sell-herding measure. S is the composite sentiment measure. RSTD is the standard deviation of fund return. CAP is the capitalization of a given stock. TNUMBER is the number of funds trading a given stock. PR is the ratio of mutual fund purchases minus mutual fund redemption in a given month to net assets of the fund. PB is the ratio of market value to book equity at the beginning of the trading quarter. ***, ** and * indicate significant at the 0.01, 0.05 and 0.10 level, respectively.

4.2 Herding comparisons between high and low sentiment sub-samples

The bivariate correlation results in Table 3 indicate that S_{t-1} relates positively with HM and SHM. Fund managers show a tendency to herd in trading stocks with high investor sentiment, especially on the sell-side. To further support this view, Table 4 presents univariate comparisons of herding between high and low S_{t-1} sub-samples. High (low) S_{t-1} sub-sample refers to observations with S_{t-1} greater (less) than the median S_{t-1} of the full sample. Relative to low S_{t-1} stocks, stocks with high S_{t-1} also have higher mean and median HM (t=5.086 and Z=3.475 respectively). Similarly, both the mean and median SHM for high S_{t-1} subsample are significantly higher than the figures for low S_{t-1} subsample (t=5.755 and Z=3.969 respectively). However, neither the mean nor median of BHM for high S_{t-1} subgroup is significantly different from that of low S_{t-1} subgroup (t=1.419 and Z=1.160 respectively). By and large, the observed fund herding behavior in high sentiment stocks mostly comes from selling herding rather than from buying herding. Consequently, Table 4 provides evidence consistent with the correlation coefficient findings in Table 3, suggesting that fund herds form more often on the sell-side than on the buy-side in stocks with high past investor sentiment.

		HM	BHM	SHM
High St-1	Mean	0.017	0.018	0.013
	Std Dev.	0.053	0.066	0.052
	Median	0.004	0.004	0.003
Low St-1	Mean	0.014	0.017	0.010
	Std Dev.	0.051	0.058	0.039
	Median	0.002	0.003	0.001
t value for the difference		5.086 (<.001)***	1.419 (0.156)	5.755 (<.001)***
Wilcoxon Z the difference		3.475 (<.001)***	1.160 (0.246)	3.969 (<.001)***

Table 4. Herding Comparisons between High and Low Sentiment Sub-samples

Note. The sample contains 770 U.S. mutual funds' holding records on 527 stocks from 2001 to 2005, including 31,093 (17,095 buy-herding and 13,998 sell-herding) stock-month observations. S_{t-1} is the sentiment index in month t-1. High (low) S_{t-1} sub-sample refers to observations with S_{t-1} greater (less) than the median S_{t-1} of the full sample. HM is the aggregate herding measure. BHM is the buy-herding measure. SHM is the sell-herding measure. The test statistics are heteroskedastic t-tests of equal means and non-parametric Wilcoxon Z-values of equal medians comparing high sentiment sub-sample with low sentiment sub-sample. Figures in parentheses are p values. *** indicates significant at the 0.01 level.

4.3 The effect of prior sentiment on fund herding

Table 5 reports the regression results for different fund herding measures: aggregate herding (HM), buy-herding (BHM) and sell-herding (SHM). The t-statistics in Table 5 are based on White's heteroskedasticity consistent estimator for standard errors. The VIFs for all independent variables are smaller than 3, indicating that collinearity does not pose an issue in model estimation (Kennedy, 2003).

Expected Sign		Dependent Variable		
		HM	BHM	SHM
		Estimated Coefficient	Estimated Coefficient	Estimated Coefficient
Intercept		0.0865	0.1514	0.1025
		(0.9307)	(1.1844)	(1.0626)
St-1	+	0.0003	0.0002	0.0003
		(2.2866)**	(1.1715)	(1.9429)*
		[1.4640]	[1.3662]	[1.3817]
RSTDt-1	+	0.2461	0.1707	0.4538
		(6.5192)***	(1.8334)*	(7.2789)***
		[1.2206]	[1.2188]	[1.4552]
CAPt	-			
		-2.4660E-10	-3.6311E-10	-7.2224E-11
		(-5.8024)***	(-6.6908)***	(-4.4602)***
		[2.5679]	[2.4828]	[2.5714]
TNUMBERt	+			
		6.7002E-5	8.7430E-5	3.2909E-5
		(6.8879)***	(3.3417)***	(2.8771)***
		[2.3489]	[2.1701]	[2.4336]
PRt	+/+/-	0.0322	0.0445	-0.0103
		(0.9796)	(3.8710)***	(-2.0621)**
		[1.1518]	[1.1440]	[1.2581]
PBt-1	+	1.6133E-6	8.8284E-7	2.7620E-5
		(0.3765)	(0.1917)	(0.9920)
		[1.0104]	[1.0399]	[1.0530]
N		31093	17095	13998
Adj. R-sq		0.0240	0.0129	0.0218
Prob. (F > F*)		<0.0001	<0.0001	<0.0001

Table 5. The Effect of Investor Sentiment on Fund Herding

Note. The sample is the 770 U.S. mutual funds' holding records on 527 stocks from 2001 to 2005, including 31,093 (17,095 buy-herding and 13,998 sell-herding) observations. HM is the aggregate herding measure. BHM is the buy-herding measure. SHM is the sell-herding measure. S is the composite sentiment measure. RSTD is the standard deviation of fund return. CAP is the capitalization of a given stock. TNUMBER is the number of funds trade in a given stock. PR is the dollar amount of purchase minus the dollar amount of redemption by mutual funds in a given month, scaled by net assets of the fund. PB is the ratio of market value to book equity at the beginning of the trading quarter. An industry dummy (equal to one for stocks of high-tech firms, zero otherwise) and four yearly dummies (2001 as the base year) are included in each model to control the possible industry and macroeconomic effects on herding behavior (not shown in the table). The t-statistics are based on White's heteroskedasticity consistent estimator for standard errors. t values are in ( ); VIFs are in [ ]; ***, ** and * indicate significant at the 0.01, 0.05 and 0.10 level, respectively.

For factors affecting HM, the concern of this study is the composite sentiment measure (S_{t-1}). S_{t-1} has a positive and significant coefficient (t=2.2866), which is consistent with the notion that funds trade in herd more frequently on stocks with prior high sentiment. S_{t-1} positively relates to SHM (t=1.9429) but insignificantly relates to BHM (t=1.1715), revealing that the positive effect of S_{t-1} on HM mostly comes from the sell-side herding rather than the buy-side herding. These findings indicate that fund managers share an aversion to stocks with high past optimistic sentiment. Overall, the evidence again corroborates the results in Table 3 and Table 4 and provides stronger support for the sentiment countering argument that institutional investors have an inclination to counter optimistic investor sentiment. Also, the evidence implicates the use of negative-feedback strategies by mutual funds as an important source of herding.

Of the control variables, the deviation of fund returns in previous month ($RSTDt$-1) has a positive effect on all three herding measures (t=6.5192, 1.8334 and 7.2789 respectively), consistent with Bikhchandani et al.'s (1992) reputational risk argument. The stock capitalization ($CAPt$) negatively relates with all three herding measures (t=-5.8024, -6.6908 and -4.4602 respectively), showing that herding is more pronounced in small stocks. This result is consistent with the various herding theories where fund managers may receive lower precision information from these firms and are more likely to ignore this information if consensus opinion is different, or fund managers may share an aversion to holding small stocks because these stocks are less liquid (Lakonishok et al., 1992; Wermers, 1999; Choe et al., 1999). The positive effect of the number of funds trading a stock ($TNUMBER_t$) on all three herding measures (t=6.8879, 3.3417 and 2.8771 respectively) is consistent with the findings of Wermers (1999) and Wylie (2005). The net mutual fund redemption (PR_t) positively correlates with BHM (t=3.8710) and negatively correlates with SHM (t=-2.0621), similar to the findings in Edelen (1999) and Gallagher and Jarnecic (2004). Finally, the coefficient of market-to-book ratio is not significant in all three models, suggesting that the growth orientation of stocks does not play a role in explaining the level of fund managers herding.

4.4 Sensitivity tests

To assess the robustness of the results, the study includes four sensitivity tests : (1) re-estimating Table 4 and Table 5 for sub-periods "2001-2002" (a bear market) and "2003-2005" (a bull market) separately; (2) excluding 5% of upper outliers to alleviate the effect of extreme BHM and SHM and this process reduces the sample size to 16240 buy-herding and 13298 sell-herding observations; (3) replacing the trinomial-distribution herding measure with the popular binomial- distribution herding measure (Lakonishok et al., 1992); and (4) replacing the investor sentiment measure (St-1) in Table 4 and Table 5 with a two-period lag measure (St-2). The results (not reported) of these tests are qualitatively similar to those reported.

5. Conclusions

This paper assesses the relation between investor sentiment and the extent to which fund managers herd in their trades of stocks. The study applies the Lakonishok et al.'s (1992) measure and Wylie's (2005) trinomial-distribution approach to gauge fund manager herding, and uses the principal component analysis as the means of extracting the composite unobserved sentiment measure from ten market weather indicators that can be categorized into three groups: individual stock sentiment indicator, overall market sentiment indicator, and fund sentiment indicator.

The empirical results suggest that investor sentiment plays a significant role in explaining mutual fund herding cross-sectionally, especially on the sell-side. Fund managers show a stronger tendency to herd out of stocks with high prior investor sentiment than to herd into stocks with high prior sentiment. In other words, managers herd sell because, at least in part, they observe and counteract the optimistic sentiment. This finding is consistent with the funds sharing an aversion to stocks that have previously demonstrated high optimistic sentiment, supporting the sentiment countering hypothesis. The results also provide support for the informational cascade theory where managers herd because of analyzing the same sentiment-related indicators.

6. References

Baker, M., Stein, J.C., 2004. Market liquidity as a sentiment indicator. Journal of Financial Markets 7(3), 271-299.

Baker, M., Wurgler, J., 2006. Investor sentiment and the cross-section of stock returns. Journal of Finance 61(4), 1645-1680.

Bala, V., Goyal, S., 1998. Learning from neighbors. Review of Economic Studies 65(3), 595-621.

Banerjee, A., 1992. A simple model of herd behavior. Quarterly Journal of Economics 107(3), 797-817.

Barberis, N.C., Shleifer, A., 2003. Style investing. Journal of Financial Economics 68(2), 161-199.

Bikhchandani, S., Hirshleifer, D., Welch, I., 1992. A theory of fads, fashion, custom, and cultural change as informational cascades. Journal of Political Economy 100(5), 992-1026.

Block, S.B., French, D.W., Maberly, E.D., 2000. The pattern of intraday portfolio management decisions: a case study of intraday security return patterns. Journal of Business Research 50(3), 321-326.

Brandenburger, A., Polak, B., 1996. When managers cover their posteriors: Making the decisions the market wants to see. Rand Journal of Economics 27(3), 523-541.

Brown, G.W., Cliff, M.T., 2004. Investor sentiment and the near-term stock market. Journal of Empirical Finance 11(1), 1-27.

Choe, H., Kho, B.C., Stulz, R.M., 1999. Do foreign investor destabilize stock markets? The Korean experience in 1997. Journal of Financial Economics 54(2), 227-264.

De Bondt, W.F., 1993. Betting on trends: Intuitive forecasts of financial risk and return. International Journal of Forecasting 9(3), 355-371.

De Long, J.B., Shlerter, A., Summers, L.H., Waldmann, R.J., 1990. Noise trader risk in financial markets. Journal of Political Economy 98(4), 703-738.

Del Guercio, D., 1996. The distorting effect of the prudent-man laws on institutional equity investments. Journal of Financial Economics 40(1), 31-62.

Edelen, R.M., 1999. Investor flows and the assessed performance of open end fund managers. Journal of Financial Economics 53(3), 439-466.

Falkenstein, E.G., 1996. Preferences for stock characteristics as revealed by mutual fund portfolio holdings. Journal of Finance 51(1), 111-135.

Fisher, K.L., Statman, M., 2000. Investor sentiment and stock returns. Financial Analysts Journal 56(2), 16-23.

Froot, K.A., Scharfstein, D.S., Stein, J.C., 1992. Herd on the street: Informational inefficiencies in a market with short-term speculation. Journal of Finance 47(4), 1461-1484.

Fung, W., Hsieh, D.A., 1999. A primer on hedge funds. Journal of Empirical Finance 6(3), 309-331.

Gallagher, D.R., Jarnecic, E., 2004. International equity funds, performance, and investor flows: Australian evidence. Journal of Multinational Financial Management 14(1), 81-95.

Gompers, P., Metrick, A., 2001. How are large institutions different from other investors? Quarterly Journal of Economics 16(1), 229-259.

Grinblatt, M., Titman, S., Wermers, R., 1995. Momentum investment strategies, portfolio performance, and herding: A study of mutual fund behavior. American Economic Review 85(5), 1088-1105.

Hirschleifer, D., Subrahmanyam, A., Titman, S., 1994. Security analysis and trading patterns: When some investors receive information before others. Journal of Finance 49(5), 1665-1698.

Jegadeesh, N., Titman, S., 1993. Returns to buying winners and selling losers: Implications for stock market efficiency. Journal of Finance 48(1), 65-91.

Kennedy, P., 2003. A guide to econometrics. Cambridge, MA: MIT Press.

Klemkosky, R.C., 1977. The impact and efficiency of institutional net trading imbalances. Journal of Finance 32(1), 79-86.

Kyrolainen, P., Perttunen, J., 2003. Investors' activity and trading behavior. Working paper, University of Oulu, Finland.

Lakonishok, J., Shleifer, A., Vishny, R.W., 1992. The impact of institutional trading on stock prices. Journal of Financial Economics 32(1), 23-43.

Lee, C., Shleifer, A., Thaler, R., 1991. Investor sentiment and the closed-end fund puzzle. Journal of Finance 46(1), 75-109.

Neal, R., Wheatley, S.M., 1998. Do measures of investor sentiment predict returns? Journal of Financial and Quantitative Analysis 33(4), 523-548.

Nofsinger, J.R., Sias, R.W., 1999. Herding and feedback trading by institutional and individual investors. Journal of Finance 54(6), 2263-2295.

Payne, T.H., Prather, L., Bertin, W., 1999. Value creation and determinants of equity fund performance. Journal of Business Research 45(1), 69-74.

Pound, J., Shiller, R.J., 1989. Survey evidence on diffusion of interest and information among investors. Journal of Economic Behavior and Organization 12(1), 47.67.

Prendergast, C., 1993. A theory of yes men. American Economic Review 83(4), 757-770.

Scharfstein, D., Stein, J., 1990. Herd behavior and investment. American Economic Review 80(3), 465-479.

Shleifer, A., 2000. Inefficient markets: An introduction to behavioral finance. Oxford University Press, Oxford.

Trueman, B., 1994. Analyst forecasts and herding behavior. Review of Financial Studies 7(1), 97-124.

Walter, A., Weber, F.M., 2006. Herding in the German mutual fund industry. European Financial Management 12(3), 375-406.

Wermers, R., 1997. Momentum investment strategies of mutual funds, performance persistence, and survivorship bias. Working paper, University of Colorado.

Wermers, R., 1999. Mutual fund herding and the impact on stock price. Journal of Finance 54(2), 581-622.

White, H., 1980. A heteroskedasticity-consistent covariance matrix estimator and a direct test for heteroskedasticity. Econometrica 48(1), 817-838.

Wylie, S., 2005. Fund manager herding: a test of the accuracy of empirical results using U.K. data. Journal of Business 78(1), 381-403.

Zwiebel, J., 1995. Corporate conservatism and relative compensation. Journal of Political Economy 103(1), 1-25.

4

Chemometrics (PCA) in Pharmaceutics: Tablet Development, Manufacturing and Quality Assurance

Ingunn Tho[1] and Annette Bauer-Brandl[2]
[1]University of Tromsø
[2]University of Southern Denmark
[1]Norway
[2]Denmark

1. Introduction

Pharmaceutical tablets are subject to special regulations regarding their quality. Obvious examples are requirements for tablet performance in terms of pharmacological effect, which is closely connected to the uniformity of drug substance content in each tablet, to the disintegration properties of the tablet into smaller particles after intake, and to the rate of dissolution of the drug substance from the tablet or particles (drug release). For each product on the market, a set of quality criteria and their specific limits are defined by the relevant regulatory health authorities, e.g. Food and Drug Administration (FDA) in the USA, or European Medicines Agency (EMA) for Europe.

Following the steps of the development of a new tablet, we distinguish six main phases:

a. design and synthesis of the new pharmaceutically active drug substance (active pharmaceutical ingredient, API),
b. characterisation and quality assurance of the bulk raw materials, i.e. the API and additives (excipients),
c. optimisation of the composition of the formulation with respect to API and excipients,
d. scale-up to larger throughput (industrial scale),
e. optimisation and validation of the processing steps for the manufacturing; monitoring processing steps during manufacture,
f. quality assurance of the finished product, the tablet.

Tablets have been produced for more than 100 years in large quantities, and their quality has always been an issue; therefore the technical factors that influence tablet properties have been studied extensively. It has immediately become obvious that the number of factors is large and that there is a lot of interaction between them. Therefore, pharmaceutical development in general, and tablet manufacture in particular, have been one of the first areas where factor analysis and Design of Experiments (DoE) have been introduced.

The current international regulations for pharmaceutical products, namely the ICH (International Conference of Harmonisation) guidelines (ICH Q8) including Quality by

Design (QbD) approaches together with Process Analytical Technology (PAT) initiated by the FDA in 2004 have drawn even more attention to the use of state-of-the-art science and technology (PAT-Initiative, 2004; Guidance for industry, 2004). Improvement of the manufacturing of drug preparations is thereby initiated with the aim to replace end-product control of random samples by real-time quality control of every single item.

It is nowadays a prerequisite for the registration of a new tablet preparation to show how the use of systematic studies during the development phase has led to a good understanding of the processes, how rational decisions have been made during the development process with regard to an optimum product, how the manufacturing processes can be controlled within set limits, and to estimate the robustness of the processes, i.e. what a certain variability of any ingoing parameters means for the properties of the final product. Therefore, it is very common to integrate factor analysis and multivariate projection methods such as PCA in all stages of pharmaceutical development and manufacture in general. However, tablets are a particularly prominent example thereof, because all stages from the preparation of the bulk powder, granulation, compression and coating widely influence the properties of the final marketed product. Systematic studies within a reasonable design spaces and a good scientific understanding of the processes reduce uncertainty and create the basis of flexible and risk-based regulatory decisions. Therefore, chemometric methods, such as DoE and multivariate analysis (MVA) are – if nothing else of necessity - frequently and extensively used in pharmaceutical product and process development.

This chapter focuses exclusively on fundamental studies for the rational development of tablet formulations of general interest. Therefore, it will restrict to examples of commonly used excipients and to simple manufacturing methods. The interested reader is addressed to specialized textbooks for further reading about production technologies and quality assurance (e.g. Bauer-Brandl & Ritschel, 2012; Gad, 2008).

2. Definitions and background information

2.1 Pharmaceutical tablets

Tablets contain both the active pharmaceutical ingredient, API, for obvious reasons, and different types of excipients to generate both manufacturability and product performance. In a standard procedure, both types of powders (API and excipients) are homogeneously mixed and granulated, the granules finally mixed with more excipients (glidants, lubricants, etc), upon which the final blend is compressed into tablets. The tablets will, in contact with water or fluids of the gastrointestinal tract, disintegrate into the granule/powder again, before the drug substance is released. Figure 1 depicts the basic steps of production of a standard tablet and processes upon contact with water or gastrointestinal fluids.

2.2 Quality of tablets

For pharmaceutical tablets, quality has many aspects: We have to distinguish between the tablets as such, and the drug product, which includes the packaging and the product description. Table 1 lists some of the most important quality criteria for a tablet product. Any of these may be used as a quality attribute for optimisation of a tablet product.

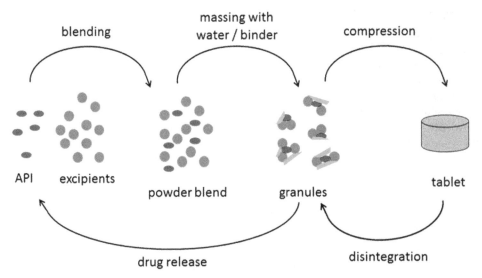

Fig. 1. Schematic drawing of the basic production steps of a tablet (upper arrows), and the disintegration and release of drug from the tablet (lower arrows)

Appearance of the tablets	Composition	Properties of the tablets	Properties of the drug product
Shape	Identity of drug substance(s)	Mechanical strength	Outer package identity
Dimensions: Height, Diameter	Additives (excipients)	Ease to break	Label identity
Engravings (Symbols, numbers, etc)	Quantitative drug content per tablet	Disintegration behaviour in water	Leaflet identity
Colour	Total mass per tablet	Drug release rates in different media	Tamper-proof inner package
Surface appearance (shiny, smooth)	Coating composition	(mouth feeling)	Safe storage conditions
No specks, no spots	Coating thickness	No capping, no layering	No sign of tampering
No cracks, no chipping			No defect

Table 1. Some important quality criteria of tablets as a drug product

For the marketing of a drug product, all the specifications need to be authorized by the respective local health authorities. Thereby, the actual values of quality requirements are defined for each single product individually, depending on the significance of the different criteria for this particular product and on special rules in the respective country.

2.3 Tablet development

The development of a new tablet formulation and production process takes its start in the properties of the active ingredient(s). For formulation (composition) development, it is of importance to choose the most suitable set of excipients to mix with the APIs to obtain a tablet product of certain quality attributes. In process (handling) development, including the choice of the processing method, all the processing parameters (such as temperatures, sequence of addition of excipients, mixing conditions, duration of the processing steps etc.) contribute to rather complex interrelated relationships between critical process parameters and quality criteria. The entire production process is typically divided into several steps, each of which is individually optimised.

Figure 2 shows general principles of how to control step-wise production. Choice of production method, optimisation of production steps, and control schemes have traditionally been handled by practical experience and by empirical development methods.

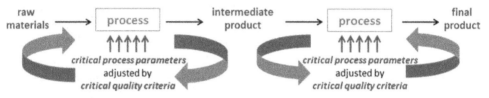

Fig. 2. Schematic drawing of production step control

2.4 A Brief overview of processes and parameters involved

In the following the elaboration will be restricted to basic tablets produced by a standard technique. Table 2 shows the impact of the main production steps on the quality of standard tablets. A schematic depiction of the sequential arrangement of some commonly used processing steps for standard tablets is given in Figure 3 together with a selection of related important processing parameters. Furthermore, monitoring of some machine parameters is also included.

2.5 The combination of DoE and MVA

Statistical experimental design, also called design of experiments (DoE), is a well-established concept for planning and execution of informative experiments. DoE can be used in many applications connected to tablet development and manufacture. One may, for instance, use designs from the *fractional factorial* or *central composite* design families, when addressing problems in which the influence of process parameters (temperature, pressure, time, etc.) on product properties are monitored and optimized. Another main type of DoE application concerns the preparation and modification of mixtures or formulations, where the challenge is that the parameters are not independent but components add up to 100%. This involves the use of *mixture designs* for changing composition and exploring how such changes will affect the properties of the formulation (Eriksson et al., 1998). Also designs in which both process and mixture factors are varied simultaneously are frequently seen. For tablet development and manufacture, one or several of the process parameters and/or critical quality attributes listed above, are likely to be used as optimization goals.

Tablet Property	Raw materials	Granulation	Drying	Final blending	Compression
Dimensions	++	+	+	+	++
Appearance	+	+	++	+	++
Total mass per tablet	+	+	++	++	-
Drug content (assay)	+	+	+	++	-
Mechanical strength	++	++	++	++	++
Disintegration	++	++	+	++	++
Drug release rate	++	++	+	++	++
Stability; degradation	+	-	++	-	-
Microbiological quality	++	-	++	-	-

+: important; ++: very important; -: important in special cases only

Table 2. Impact of Processing Steps on Tablet Quality

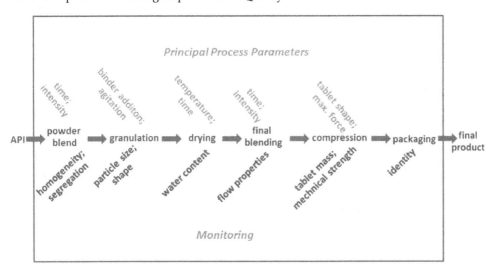

Fig. 3. A schematic view of the course of main processing steps, processing parameters and monitoring features of tablet production

Multivariate projection methods, such as principal component analysis (PCA) and partial least square (PLS) regression, are combined with DoE in formulation or process development for screening of a large number of influence parameters (fractionated factorial design) and further optimization of the important parameters, to study their interactions and possible non-linear behaviour (central composite design, Box Behnken, D-optimal, mixture designs).

3. PCA in exploring new molecular entities

QSAR and QSPR respectively (quantitative structure activity/property relationship) interlink biological effects or physical properties to the molecular structure of the respective substance. The best known example is the design and synthesis of new drug molecular entities in medicinal chemistry where the aim is that the drug molecule (API) interacts with the molecular structures at the site of action (e.g. receptors, enzymes). Another important step for a drug substance to act as a useful remedy is the uptake of such molecules into the body, which is a prerequisite to reach the receptor – has led to systematic PCA based studies of structure permeability relationships (e.g. intestinal barriers for the uptaker of oral drugs, or transport though the blood brain barrier; BBB partitioning).

Similar PCA methods have also been applied in formulation development by means of calculated quantum chemical descriptors for excipients (see Section 4.1.1.).

4. PCA in formulation development and processing

For the development of the formulation (composition) and the process (handling), all the experimental parameters during the entire manufacturing (such as chemical composition, processing steps, and processing conditions) contribute to the properties of the final product. In the following, examples of recent studies are briefly discussed, organized according to separate steps, although the entire flow of processing steps affects the product properties.

4.1 Excipient choice

As a rule, excipients have multiple functionality depending on the types of (other) materials present in the formulation, composition of the blends, and processes used. Well-known examples are filler-binders and lubricant-glidants.

A set of basic studies including four groups of excipients (filler, binder, disintegrant, lubricant) with several examples for each, in total almost one hundred excipients, were screened with respect to tablet properties (good mechanical strength and short disintegration time) (Gabrielsson et al., 2000; 2004). DoE combined with PCA and PLS were used in screening and optimisation: PCA provides a general overview over relationships and PLS models quantify the relationships between excipient properties and responses (tablet properties). Such mapping of excipients can be useful for selection of the appropriate material for a new formulation.

4.1.1 Examples of functional polymers as excipients

Microcrystalline cellulose (MCC) is a frequently used excipient with filler/binder function. It is derived from native cellulose through hydrolysation and purification. Using DoE and stepwise multiple regression analysis, particle properties and the functionality of different MCC products can be quantitatively related to the hydrolysis conditions (Wu et al., 2001).

As another example the formulation of pectin pelletisation masses using different granulation liquids is shown in the form of a PCA score plot mapping the experimental area. Correlation between properties of the masses and the respective additives was found using quantum chemical molecular descriptors (QSAR) of the additives. Partial least square regression methods, mainly PLS-2 models, were used to quantify the effects (Tho et al., 2002). Based on the models, a rational choice of the composition of the formulation can be made.

4.2 Mixing / blending

Homogeneous powder blends are a prerequisite of the predefined dose of API per tablet. In optimized blending processes, homogeneity of the bulk is achieved.

However, segregation tendencies – becoming significant even for initially homogeneous bulk powder blends when these materials are transported - are even more important to consider. The effect of powder properties on segregation was studied, and PCA models developed to connect material properties to segregation tendency. A comparison of the multivariate approach and univariate approaches reveals that the latter leads to incomplete models (Xie et al., (2008)).

4.3 Particle size and powder flow

Multivariate latent variable methods have also been applied to mixture modelling of pharmaceutical powders. Particle size distribution was shown to predict powder flow (Mullarney and Leyva, 2009). Particle size distribution along with particle shape information has also been used to predict bulk packing and flow behavior (Sandler & Wilson, 2010).

4.4 Granulation

Wet granulation is classically done in high shear mixers or in fluidized bed equipment, or in a recent approach as a continuous process.

A basic investigation of general interest into the fluid bed granulation process is described by Dias and Pinto (2002), who used DoE and cluster analysis to find the relevant processing parameters and relations to product properties.

Another basic study of the fluid bed granulation process is described by Otsaka et al. (2011). PCA was first used to describe the relationships between a set of granule properties and to identify the properties that produce optimum tablet properties. In a second step, regression modelling was used to optimize the granulation conditions.

For wet granulation in high shear mixers, there are two different possibilities: The first one is a separate wet massing step and subsequent drying in another machine. The alternative is a single-pot set-up, where the drying step can be conducted in the same piece of equipment. Giry et al. compare these two processes (Giry et al., 2009) with respect to product properties using a set of formulations. This is a typical DoE and PCA application. Both processes led to products with only slight differences in properties. However, the robustness of the processes was different.

An alternative to wet granulation methods is dry granulation, basically by compression of dry powder blends to make granules. A dry granulation process conditions was studied within a PLS framework to predict selected granulation properties (Soh et al. 2008).

4.5 Drying

In-line monitoring of the water content is essential in order to find the end point of the drying step for optimum water content of granules. NIR (near infrared) spectroscopy is particularly sensitive to water content, and using PCA/PLS models enable prediction of the actual water content. In addition, the NIR spectra are also sensitive to other granule properties such as particle size. However, in bulk material of a wide particle size, the actual water content of the granules varies significantly with size. Nieuwmeyer et al. (2007) developed a method for characterization of both water content and granule size during fluid bed drying based on NIR spectroscopy.

4.6 Final blend; lubrication

The final blend is typically produced by blending solid lubricant into the granules. The problem with too long mixing times ("overmixing") is that lubricant particles scale off and cover too much of the granule surface, which leads to weak tablets. As an example to monitor the blending profile (the degree of lubricant coverage) over time, and correlate this to tablet hardness, NIR was used (Otsuka & Yamane, 2006). Principal component regression was employed to model and predict the hardness of the tablets.

4.7 Tablet compression

Compression of the materials into tablets is a complex processing step, which is not yet fully understood. Amongst others, influences of particle size, shape, and structure on the deformation mechanism are difficult to separate and quantify. It has been shown that it is possible to derive basic material deformation characteristics during the compression step from in-line processing data using model materials (Klevan et al., 2010). Due to the complexity of the deformation of powder beds and the large number of parameters that may be regarded, screening of such in-line derived parameters for the most useful set of parameters is necessary in order to avoid over-determined functions (Andersen at al., 2006). Examples are discussed using a set of commonly used excipients of most diverse properties within the useful materials (Haware et al., 2009a; Roopwani & Buckner, 2011). The effect of particle engineering with regard to the tablet properties, such as tablet tensile strength, studied by PLS models (Otsuka & Yamane, 2006; Haware et al., 2010) using cross validation (jack-knifing). Prediction of tablet mechanical properties using PLS model can be made from NIR spectra of the blends (Otsuka & Yamane, 2009) or from in-line compression parameters derived on examples of very similar properties (Haware et al., 2009b).

Figure 4 shows a PCA biplot as an example how particle properties of different commercially available lactose qualities spread out in a design space. The data together with in-line compression parameters were used to predict the mechanical strength of tablets made with different blending ratios (formulations) and under different experimental conditions of compression (Haware et al., 2009b).

4.8 Combination of processes

Polizzi and García-Munoz (2011) proposes a quantitative approach to simultaneously predict mechanical properties of particles, powders, and compacts of a pharmaceutical blend, based on the raw materials. They used a two-step, multivariate modeling approach created using historical physical data of APIs, excipients, and multi-component blends. The physical properties for each individual component were first transformed using a PCA technique to place them in a multivariate design space and capture property correlations. The scores from these PCA models were then weighted by the blending ratios prior to PLS regression versus actual measured blend properties. This method produced a complete prediction of all the material properties simultaneously which was shown to be superior to the prediction performance observed when applying linear ideal-mixing.

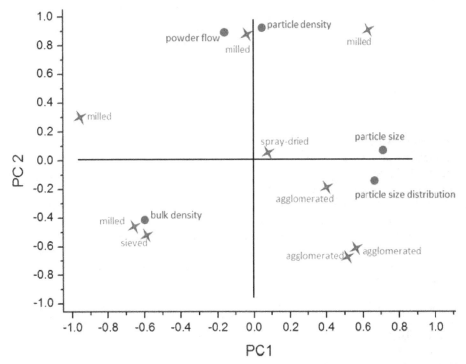

Fig. 4. Example of a PCA biplot showing powder characteristics of commercially available lactose grades spread out in the design space (modified after Haware et al. 2009b).

5. PCA in quality assurance of tablets

5.1 Spectroscopic methods

Molecular vibrational spectroscopy techniques, such as infrared (IR), near infrared (NIR) and Raman, are characterisation methods that have been applied to monitor both physical and chemical phenomena occurring during the processing as well as for off-line characterisation of raw materials and end-products. These techniques produce data of high dimensionality, since each sample is described with hundreds or even thousands of variables (wavelengths). Multivariate projection techniques, such as PCA are frequently combined with spectroscopic methods to enable detection of multivariate relationships between different variables such as raw materials, process conditions, and end products. Also well-known spectroscopic methods that are easier to interpret, e.g. UV (ultraviolet) spectroscopy and X-Ray diffraction, have been suggested to benefit from the use of PCA.

Both NIR and Raman spectroscopic measurements can be done in a very fast (within seconds) and non-destructive way, making them suitable tools for real-time process monitoring. They are frequently used for in-line monitoring of processes, e.g. particle size growing during granulation or the water content in the granule (Räsänen & Sandler, 2007; Rantanen 2007). NIR is also used for monitoring the final quality of the single tablets, because it is possible to quantify the active ingredient in a none-destructive manner (Tabasi et al., 2008), and can be applied to indentify counterfeit drugs (Rodionova et al., 2005). Another example is the prediction of tablet hardness based on NIR spectra of powders with lubricants (Otsuka & Yamane, 2006, 2009). For further applications of vibrational spectroscopy and chemometric methods in pharmaceutical processes, the reader is referred to the reviews by Gendrin et al. (2008), De Beer et al. (2011) and Rajalahti & Kvalheim (2011).

5.2 PCA in drug release

Sande and Dyrstad (2002) used PCA/PCR for parameterisation of kinetic drug release profiles, and showed that the combination of PCA and SIMCA was useful for classification of formulation variables based on the entire profiles. Also multi-way principal component analysis (MPCA) has been used to study entire tablet dissolution profiles and detection of shifts upon accelerated stability (Huang et al., 2011). Korhonen et al. (2005) calculated physicochemical descriptors (using VolSurf software) of different model drugs and correlated those to the respective dissolution profiles from matrix tablets. Again, the entire dissolution profiles without any parametric curve fitting were used in the models. This is a great advantageous, since the contribution of different variables tends to change in the course of the drug release event due to the dynamic nature of drug release from the matrix tablets.

Another example of the non-parametric curve description of drug release from a sustained release matrix is the approach suggested by Kikuchi and Takayama (2010). They use non-parametric descriptors and non-parametric differences between release curves for optimization (24 h linear release) and prediction. Surface plots are used for prediction of composition of the matrix tablets.

6. Monitoring and sensors

Combination of instrumentation and multivariate analysis provides powerful tools for effective process monitoring and control enabling detection of multivariate relationships between different variables such as raw materials, process conditions, and end products. Multivariate methods play an important role in process understanding, multivariate statistical process control, default detection and diagnosis, process control and process scale-up.

Continuous monitoring of process parameters allows continuous regulation and adjustment. Given rational development of the parameters as a background, a statistical process control is possible in order control the variation in a process within specification limits.

Furthermore, non-statistical variation caused by a specific event can be identified and the special causes thereof studied, with a possibility to eliminate these for the future.

In the case of tableting, continuous monitoring of the maximum compression force for each single tablet has been a standard in industrial production for more than 30 years. The force values are used as a measure for the individual tablet mass (and given homogeneity of the tableting material, also the dosing of each tablet). It is possible to sort out single bad tablets from the production line. A feedback loop needs to readjust the correlation between force and dosing based on sample analysis. By this, each tablet is held within the specification limits. Further development on tablet machine instrumentation includes measuring forces and displacements at different machine parts, and acoustic sensors in order to catch effects connected to lubrication problems.

Statistical Process Control (SPS) is the best way to take advantage of real-time monitoring. Based on the experience from process development, upper and lower warning and control limits can be set, where predefined action is taken: further trend analysis, actual changing of parameters, searching for the cause of an outlier. Figure 5 illustrates how SPS may be implemented in process monitoring. Trending analysis and feedback loops are necessary features to readjust these limits if necessary. In cases where secondary measures are used, reassuring the calibration with other measures is frequently conducted on a regular basis. A typical example is the relationship between compression force and mass of tablets. This is directly connected to the bulk density of the tableting material, which may change during the course of the actual tableting process. In many cases increased density is observed over time due to bulk material transport in the machine combined with machine vibrations. The mass of the tablets therefore needs to actually be checked on a balance.

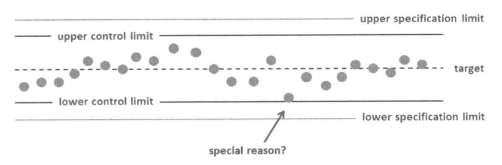

Fig. 5. Principles of Statistical Process Control (SPS) based on process parameter monitoring

Frequently used sensors based on NIR or Raman spectroscopy identify the chemical or physical variation between intermediates and products, particularly in the case of large numbers of samples:

- particle size distribution of powders and granules,
- granulation process monitoring,
- powder mixing optimization,
- scale-up of powder mixing,
- tablet properties (e.g. drug content).

These methods can be used to collect all batch characteristics, model all batches, detect outliers, and check future batches against historic batch models (in-line monitoring). These are prerequisites to diagnose out-of-limit-batches.

7. Future of tablet manufacture development

Traditionally, pharmaceutical manufactures are batch-based production methods with sample-based end-product control. This includes a sampling procedure with all its drawbacks. The future will be in the real-time in-line monitoring during processing, in order to be able to control the processing steps rather than end-point controls on the product. This will consequently lead to continuous production processes, which can be conducted in equilibrium. Continuous granulation has been developed already for a number of processes, including fluid-bed granulation, extrusion, and roller-compaction. These days, complete continuous production lines from the powder to the tablet are being introduced to the market.

The vision is to have such continuous production all the way from API synthesis to the final drug product ready for the market. The advantage will be a controlled and stable process with less product variation compared to traditional batch-based production methods. Furthermore, sample-based post-production quality control will be unnecessary and discarding or reworking of bad batches will not happen again. The real-time-release of the product will also become possible.

8. Conclusion

Although pharmaceutical tablets have been produced for more than 100 years on an industrial scale, there are up to day a number of unresolved challenges in tablet development, manufacturing and quality assurance. The basic factors that influence the product properties are widely known. However, depending on the individual composition of the tablet, numerous factors may have different impact on product quality. Furthermore, there is commonly a large interaction between many of these factors. Depending on composition, processing conditions for the different steps and environmental conditions the properties of the final bulk material (tableting blend) can differ widely. The actual tablet compression step, again committed to numerous processing factors, in addition would have large impact on the product quality. These complex relationships open a multivariate perspective and require suitable statistical methods.

For the tableting procedure, the steps that have been studied include powder blending, granulation, particle egalisation, lubrication, compression, coating, and drug release studies. Such step-wise studies have brought light into the impact of the parameters and their interactions and increased the understanding of the respective processes.

Pharmaceutical products underlie special requirements for their quality set by the health authorities. Therefore, the pharmaceutical development has been one of the first areas where factorial analysis and PCA/PLS have even become compulsory.

Moreover, the requirements for pharmaceutical drug products are under ongoing revision. It is a general trend that the requirements set higher standards with each new revision in order to assure safety and efficacy of the products. Some of the most important recent developments are the introduction of "Quality by Design" and PAT. It is a prerequisite to understand the processes, factors and interactions for the entire production line. Figure 6 depicts how the statistical tools are related to each other, and used interconnected in tablet development, manufacture and quality assurance.

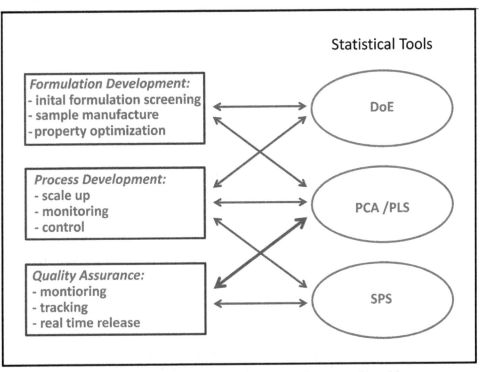

Fig. 6. A schematic overview of which statistical tools mainly are used in tablet development, manufacture and quality assurance

The development in the regulatory guidelines combined with new technology has found its expression in the efforts towards continuous production of pharmaceutical products, and in the forefront thereof the tablet manufacture due to both its relevance and complexity.

9. References

Andersen, E., Dyrstad, K., Westad, F. & Martens, H. (2006). Reducing over-optimism in variable selection by cross-model validation, *Chemometrics and Intelligent Laboratory Systems* Vol. 84 (No. 1-2): 69-74.

Bauer-Brandl A & Ritschel W.A. (2012). *Die Tablette: Handbuch der Entwicklung, Herstellung und Qualitatssicherung*, 3rd edition, Editio Cantor Verlag, Aulendorf, Germany.

De Beer, T., Burggraeve, A., Fonteyne, M., Saerens, L., Remon, J.P. & Vervaet, C. (2011). Near infrared and Raman spectroscopy for the in-process monitoring of pharmaceutical production processes, *International Journal of Pharmaceutics* Vol. 417 (No. 1-2): 32-47.

Dias, V. & Pinto, J.F. (2002). Identification of the most relevant factors that affect and reflect the Quality of Granules by application of canonical and cluster analysis, *Journal of Pharmaceutical Sciences* Vol. 91 (No. 1): 273-281.

Eriksson, L., Johansson, E. & Wikström, C. (1998). Mixture design- design generation, PLS analysis, and model usage, *Chemometrics and Intelligent Laboratory Systems* Vol. 43 (No. 1-2): 1-24.

Gabrielsson, J., Nyström, Å. & Lundstedt, T. (2000). Multivariate methods in developing an evolutionary strategy for tablet formulation, *Drug Development and Industrial Pharmacy* Vol. 26 (No. 3): 275-296.

Gabrielsson, J., Lindberg N-O., Pålsson, M., Nicklasson, F., Sjöström, M. & Lundtsedt, T. (2004). Multivariate methods in the development of a new tablet formulation; optimization and validation, *Drug Development and Industrial Pharmacy* Vol. 30 (No. 10): 1037-1049.

Gad, S.C. (ed.) (2008). *Pharmaceutical Manufacturing Handbook: Production and Processes, Section 6: Tablet Production*, Wiley-Interscience, Hoboken, New Jersey, USA, pp 879-1222.

Gendrin, C., Roggo, Y. & Collet, C. (2008). Pharmaceutical applications of vibrational chemical imaging and chemometrics: A review, *Journal of Pharmaceutical and Biomedical Analysis* Vol. 48 (No. 3-4): 533–553.

Giry, K., Viana, M., Genty, M., Louvet, F., Désiré, A., Wüthrich, P. & Chulia, D. (2009). Comparison of single pot and multiphase high shear wet granulation processes related to excipient composition, *Journal of Pharmaceutical Sciences* Vol. 98 (No. 10): 3761-3775.

Guidance for Industry (2004). PAT - A Framework for innovative pharmaceutical manufacturing and quality assurance. Available from: www.fda.gov/downloads/ Drugs/GuidanceComplianceRegulatoryInformation/Guidances/UCM070305.pdf [Last accessed 30. Oct. 2011]

Haware, R., Bauer-Brandl, A. & Tho, I. (2010). Comparative Evaluation of the Powder and Compression Properties of Various Grades and Brands of Microcrystalline Cellulose by Multivariate Methods, *Pharmaceutical Development and Technology* Vol. 15 (No. 5): 394-404.

Haware, R., Tho, I. & Bauer-Brandl, A. (2009a). Application of multivariate methods to compression behavior evaluation of directly compressible materials, *European Journal of Pharmaceutics and Biopharmaceutics* Vol. 72 (No. 1): 148-155.

Haware, R., Tho, I. & Bauer-Brandl, A. (2009b). Multivariate analysis of relationships between material properties, process parameters and tablet tensile strength for α-lactose monohydrates, *European Journal of Pharmaceutics and Biopharmaceutics* Vol. 73 (No. 3): 424-431.

Huang, J., Goolcharran, C. & Ghosh, K. (2011). A Quality by Design approach to investigate tablet dissolution shift upon accelerated stability by multivariate methods, *European Journal of Pharmaceutics and Biopharmaceutics* Vol. 78 (No. 1): 141-150.

ICH Q8 – Pharmaceutical Development, Available from: http://www.ich.org/fileadmin/ Public_Web_Site/ICH_Products/Guidelines/Quality/Q8_R1/Step4/Q8_R2_Guid eline.pdf [Last accessed 30. Oct. 2011

Kikuchi, S. & Takayama, K. (2010). Multivariate statistical approach to optimizing sustained release tablet formulations containing diltiazem HCl as a model highly water-soluble drug, *International Journal of Pharmaceutics* Vol. 386 (No. 1): 149-155.

Klevan, I., Nordstroem, J., Tho, I. & Alderborn, G., (2010). A statistical approach to evaluate the potential use of compression parameters for classification of pharmaceutical powder materials, *European Journal of Pharmaceutics and Biopharmaceutics* Vol. 75 (No. 3): 425-435.

Korhonen, O., Matero, S., Poso, A. & Ketolainen, J. (2005). Partial Least Squares Projections to latent structure analysis (PLS) in evaluation and prediciting drug release from starch acetate matrix tablets, *Journal of Pharmaceutical Sciences* Vol. 94 (No. 12): 2716-2730.

Mullarney, M.P. & Leyva, N. (2009). Modeling pharmaceutical powder-flow performance using particle-size distribution data. *Pharmaceutical Technology* Vol. 33 (No. 3): 126-134.

Nieuwmeyer, F.J.S., Damen, M., Gerich, A., Rusmini, F., van der Voort Maarschalk, K. & Vromans, H. (2007). Granule characterization during fluid bed drying by development of a near infrared method to determine water content and median granule size, *Pharmaceutical Research* Vol. 24 (No. 10): 1854-1861.

Otsuka, T., Iwao, Y., Miyagishima, A. & Itai, S. (2011). Application of PCA enables to effectively find important physical variables for optimization of fluid bed granulation conditions, *International Journal of Pharmaceutics* Vol. 409 (No. 1-2): 81-88.

Otsuka, M. & Yamane, I. (2006). Prediction of tablet hardness based on near infrared spectra of raw mixed powders by chemometrics, *Journal of Pharmaceutical Sciences* Vol. 98 (No. 7): 1425-1433.

Otsuka, M. & Yamane, I. (2009). Prediction of tablet hardness based on near infrared spectra of raw mixed powders by chemometrics: Scale-up factor of blending and tableting process, *Journal of Pharmaceutical Sciences* Vol. 98 (No. 11): 4296-4305.

Polizzi, M.A. & García-Munoz, S. (2011). A framework for in-silico formulation design using multivariate latent variable regression methods, *International Journal of Pharmaceutics* Vol. 418 (No. 2): 235-242.

Roopwani, R. & Buckner, I. (2011). Understanding deformation mechanisms during powder compaction using principal component analysis of compression data, *International Journal of Pharmaceutics* Vol. 418 (No. 2): 227-243.

PAT - Initiative (2004). Available from: http://www.fda.gov/AboutFDA/CentersOffices/CDER/ucm088828.htm [Last accessed 30. Oct. 2011]

Rajalahti, T. & Kvalheim O.M. (2011). Multivariate data analysis in pharmaceutics: A tutorial review, *International Journal of Pharmaceutics* Vol. 417 (No. 1-2): 280-290.

Rantanen, J. (2007). Process analytical applications of Raman spectroscopy, *Journal of Pharmacy and Pharmacology* Vol. 59 (No. 2): 171-177.

Rodionova, O.Y., Houmøller, L.P., Pomerantsev, A.L., Geladi, P., Burger, J., Dorofeyev, V.L. & Arzamastsev, A.P. (2005). NIR spectrometry for counterfeit drug detection. A feasibility study, *Analytica Chimica Acta* Vol. 549 (No. 1-2): 151-158.

Räsänen, E. & Sandler, N. (2007). Near infrared spectroscopy in the development of solid dosage forms, *Journal of Pharmacy and Pharmacology* Vol. 59 (No. 2): 147-159.

Sande, S.A. & Dyrstad, K. (2002). A formulation development strategy for multivariate kinetic responses, *Drug Development and Industrial Pharmacy* Vol. 28 (No. 5): 583-591.

Sandler, N. & Wilson, D. (2010). Prediction of granule packing and flow behavior based on particle size and shape analysis. *Journal of Pharmaceutical Sciences* Vol. 99 (No.): 958-968.

Soh, J.L.P., Wang, F., Boersen, N., Pinal, R., Peck, G.E., Carvajal, M.T., Cheney, J., Valthorsson, H. & Pazdan, J. (2008). Utility of multivariate analysis in modeling the effects of raw material properties and operating parameters on granule and ribbon properties prepared in roller compaction, *Drug Development and Industrial Pharmacy* Vol. 34 (No. 10): 1022-1035.

Tabasi, S.H., Fahmy, R., Bensley, D., O'Brien, C. & Hoag, S.W. (2008). Quality by design, part I: application of NIR spectroscopy to monitor tablet manufacturing process. *Journal of Pharmaceutical Sciences* Vol. 97 (No. 9): 4040-4051.

Tho, I., Anderssen, E., Dyrstad, K., Kleinebudde, P. & Sande, S.A. (2002). Quantum chemical descriptors in the formulation of pectin pellets produced by extrusion/spheronisation, *European Journal of Pharmaceutical Sciences* Vol. 16 (No. 3): 143-149.

Wu, J.-S., Ho, H.-O. & Sheu, M.-T. (2001). A statistical design to evaluate the influence of manufacturing factors on the material properties and functionalities of microcrystalline cellulose, *European Journal of Pharmaceutical Sciences* Vol. 12 (No. 4): 417-425.

Xie, L., Shen, M., Augsburger, L.L., Lyon, R.C., Khan, M.A., Hussain, A.S. & Hoag, S.W. (2008). Quality by Design: Effects of testing parameters and formulation variables on the segregation tendency of pharmaceutical powder, *Journal of Pharmaceutical Sciences* Vol. 97 (No. 10): 4485-4497.

5

Pharmacophoric Profile: Design of New Potential Drugs with PCA Analysis

Érica C. M. Nascimento and João B. L. Martins
Universidade de Brasília, LQC, Instituto de Química
Brazil

1. Introduction

Searching for the pharmacophoric profile based on the concepts of chemical and structural contribution of the receptor active sites as well as receptor ligand interactions are fundamental for the development of new potential drugs for several different diseases and body dysfunctions such as degenerative brains disorders, Alzheimer disease, Parkinson's, diabetes Mielittus, cancer and many others known sickness conditions.

Basically, a pharmacophore describes the main molecular features regarding the recognition of a ligand by a biological macromolecule. In accord to the IUPAC (International Union of Pure and Applied Chemistry) Pharmacophore can be considered as "the largest common denominator shared by a set of active molecules to specific biological target"(Wermuth et al., 1998).

A pharmacophore is defined by the pharmacophoric descriptors that are a set of electronic, structural, electrostatic and topological properties. Those descriptors can be obtained by experimental and theoretical studies. Experimental studies like X-ray crystallography, spectroscopic measurements are used to define some atomic properties of a molecule, but need a previous job – a synthesis of such molecule. Thus, in this way theoretical studies eliminate this previous job the *in silico* experiments or the theoretical studies are important tools to know a molecule before this one gets synthesized.

In modern computational chemistry, pharmacophores are used to define the essential features of one or more molecules with the same biological activity. In the past few years, it has been common that the drug discovery has a contribution from pharmacophore modeling to identify and develop new potential molecules with desired biological effect (Steindl et al., 2006).

Therefore, in order to enhance the insights to identify potential new medicinal drugs many computational strategies are widely used (Steindl et al., 2006; Tasso, 2005). Mathematics and chemical molecular modeling applied on bioinformatics are important examples of these strategies. Multivariate analysis (principal component analysis - PCA) and quantum chemistry calculations (density functional theory - DFT) are some of them, and can led to the identification of the main information required to describe the essential pharmacophore profile.

PCA has been used to find new potential molecules in a series of biological systems (Carotti et al., 2003; de Paula et al., 2009; Nascimento, et al., 2011). It is a method abundantly used in a lot of multivariate analysis (Jolliffe, 2002). This method has two main aims: to decrease variable sets in the problems of multivariate data and to select the best properties (linearly independents) that describe the system (principal components). According to Jolliffe (Jollife, 2002) the central idea of PCA analysis "is to reduce the dimensionality of a data set in which there are a large number of interrelated variables, while retaining as much as possible of the variation present in the data set. This reduction is achieved by transforming to a new set of variables, the principal components, which are uncorrelated, and which are ordered so that the first *few* retain most of the variation present in *all* of the original variables. Computation of the principal components reduces to the solution of an eigenvalue-eigenvector problem for a positive-semidefinite symmetric matrix".

Although it may seem a simple technique, PCA can be applied in a lot of problems with several variables, such as the problems involving the determination of the pharmacophoric profile of a particular class of molecules.

There are different pretreatment methods for the data analysis, e.g., mean-centering and autoscaling. This work covers the second method. The autoscaling is very similar to the mean-centering with additional paths.

First, it is calculated the mean of the data matrix X, where the columns correspond to the variables and the lines are the samples (or objects). The mean is the jth column vector of a data matrix:

$$\bar{x}_n = \frac{1}{M} \sum_{m=1}^{M} x_n \tag{1}$$

After mean-centering the standard deviation (Equation 2) for each column is calculated and the autoscaling is obtained by the division of standard deviation (Equation 3).

$$s_n^2 = \frac{1}{M-1} \sum_{m=1}^{M} (x_n - \bar{x}_n)^2 \tag{2}$$

$$\bar{x}_{ij} = \frac{x_{ij} - \bar{x}_i}{s_i} \tag{3}$$

Within the autoscaling all variables are adjusted to the same range, such that each variable has a zero mean, while the standard deviation is one. Therefore, autoscaling is important for data set where the variables have different units.

DFT is a method to investigate the electronic structure of many-body systems, like atoms and molecules (Cramer, 2004). This method can be used to calculate electronic and structural properties of molecules with biological activity like new drugs to treat a disease. Applying functional that are spatially dependent on the electronic density of multi-electronics systems.

The application of human and economic resources to improve life quality has promoted benefits and has led to the increasing on the number of diseases known as age-dependent.

Among these diseases the most frequently are the dementias, diabetes and cardiovascular disease. Among the several types of dementia existing and the most common incidents in the population over 60 years of age are Alzheimer's disease (AD), vascular dementia and Pick's disease. It is estimated that 45 million people worldwide have some kind of dementia, among these, 18 million have symptoms characteristic of Alzheimer's disease. AD is a disease that has no determined cause, and therefore their treatment is based on drugs that only treat the disease in order to remediate the cognitive functions and slow down the degenerative advance. (Sugimoto, 2002)

AD is a kind of dementia that affects neural cognitive function. Thus, one of the most widely used strategies for the treatment of this disease consists in blocking off the hydrolysis (Figure 1) of the neurotransmitter acetylcholine in cholinergic synapses of neurons, this strategy is called cholinergic hypothesis, nowadays a major target in the development of drugs that benefits and improves the chemical balance of the concentration of acetylcholine in patients with AD (Tezer, 2005; Camps, 2005; Costantino, 2008).

Some drugs act as inhibitors of AChE, among them, tacrine first drug approved by FDA for the treatment of AD (Sugimoto, 2002; Ul-haq, 2010), followed by donepezil (Sippl, 2001), rivastigmine (Sugimoto, 2002; Ul-haq, 2010) and galantamine (Sippl, 2001). Other drugs have been clinically studied and tested for use in the treatment of AD, such as physostigmine (Sippl, 2001; Ul-haq, 2010; Bartolucci, 2006). Some others are being tested and are promising candidates for the approval, including, Huperzine A (Patrick, 2005; Barak, 2005), metrifonate (Sippl, 2001) and phenserine (Sippl, 2001; Ul-haq, 2010). These drugs are indicated to treat mild to moderate stages of AD, when the patient still has independent cognitive activity.

acetylcholine　　　　**Choline**　　　　**Acetic acid**

Fig. 1. Hydrolysis of acetylcholine

Chemically, the inhibitors mentioned above have in common the inhibitory action of AChE, but have different structure and chemical nature. The geometries of some AChEIs are shown in Figure 2. Organophosphates, alkaloids and acridines, are some of the classes of drugs that inhibit acetylcholinesterase and are used in the treatment of AD. This raises the question: what are the electronic, structural, topological and chemical properties that correlate these different classes of drugs as inhibitors of the same biological target?

The brain is the most complex system of the human body and main focus of biochemical and neuromolecular studies of the twenty -first century. The research and development of drugs that act more effectively in the treatment and the study of brain's diseases mechanisms are subject to various areas of knowledge.

PCA have been used to find new potential molecules in a series of biological systems (Carotti, 2003, de Paula, 2009; de Paula, 2007; Li, 2009; Nascimento, 2011; Nascimento, 2008; Rocha, 2011; Steindl, 2005). A methodology based on PCA for the pharmacophore identification of the acetylcholinesterase inhibitors (AChEI) was recently used (de Paula, 2009; de Paula, 2007; Fang, 2011; Nascimento, 2011; Nascimento, 2008). In this chapter the electronic, structural, topological and chemical properties of studied molecules, were taken from the theoretical procedures and related to the activity by means of the multivariate analysis PCs that provides the pharmacophoric profile of these AChEIs.

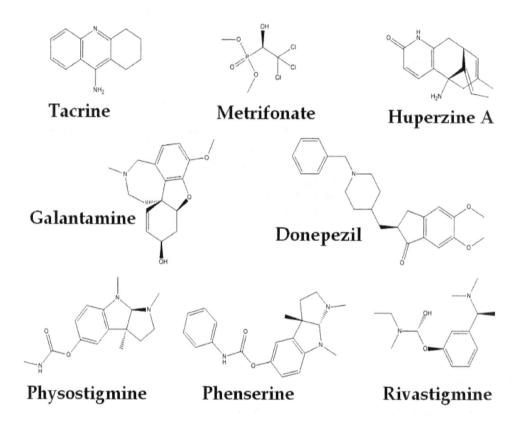

Fig. 2. Structures of some acetylcholinesterase inhibitors.

2. Mapping the pharmacophoric profile of acetylcholinesterase inhibitors

2.1 Acetylcholinesterase inhibitors, the AChEIs molecules

The cholinesterase inhibitors can be used for various purposes, as chemical warfare weapons such as sarin gas, in diseases treatment, e.g., the Alzheimer's disease treated with rivastigmine. There were used two different concepts to the rational design of inhibitors of this enzyme:

- drugs that have the molecular recognition in the catalytic site of AChE, doing preferably covalent bonds with the hydroxyl of the serine residue 200, this bond may be reversible or irreversible;
- drugs that are recognized by the to active site of the enzyme, which act by blocking the entry of the natural substrate, the neurotransmitter acetylcholine at the active site of AChE.

Some drugs that inhibit the function of the enzyme have also neuroprotective function, acting more effectively in the AD treatment. These drugs act in a negative way in the process of hyperphosphorylation of β-amyloid protein, inhibiting the breakdown of this protein and consequently the formation of amyloid plaques.

The inhibitors tacrine (THA), donepezil (E2020), rivastigmine (RIVA), galantamine (GALA), metrifonate (METRI), dichlorvos (DDVP), phenserine (PHEN), physostigmine (PHYSO), huperzine A (HUPE) and the tacrine dimer (DIMTHA) been studied theoretically by means of molecular modeling. From these studies, the mapping of pharmacophoric profile of these AChEIs was important to elucidate the correlation of such molecules as drugs to treating Alzheimer's disease.

2.2 Molecular modeling

Molecular modeling is one of the fields of theoretical chemistry that has been a great development in recent decades. The rational drug design is an area of application of molecular modeling of high relevance. Predict computationally a molecule with pharmacological activity and a drug candidate brings the expectation of reducing the search time and the cost in the molecular synthesis process and the clinical tests *in vitro* and *in vivo*.

The molecular modeling is characterized by applications of classical mechanics, quantum mechanics and stochastic methods like Monte Carlo in chemical and biochemical systems. One possibility is the prediction of spatial behavior and chemical interaction between a receptor and its biological substrate (ligand). Another is the calculation of molecular properties relevant to determining how this interaction takes place. Modeling the system receptor-ligand (RL) is possible to obtain important information about the interactions. These RL interactions establish the functionality of a given molecule. However, only the studies of interactions do not provide enough parameters to the development of new ligand.

The properties of a molecule can be obtained by theoretical calculations using classical mechanics and quantum mechanics. In the molecular mechanics approach, the classical mechanics deals with the problem without taking into account the electrons. Thus, electronic properties can only be treated by methods of electronic structure by means of quantum mechanical calculations. The main properties that describe a molecule are classified according to how it is defined spatially.

With the development of computational tools (hardware and software) it is possible today to treat complex molecular systems using approximate methods of quantum theory. Quantum mechanics considers that the energy of a molecule can be described as a sum of terms of energy resulting from electrostatic interactions between core and electrons (kinetic and potential energy). The fundamental postulate of quantum mechanics describes a wave function for each chemical system where a mathematical operator can be applied to this function resulting in observable properties of such system. Equation 4 is the focus of the study of quantum methods for the description of chemical systems and is known as the Schrödinger equation, independent of time and without relativistic corrections;

$$H\Psi = E\Psi \qquad (4)$$

where **H** is a mathematical operator that returns the energy eigenvalue of the system as an operator called the Hamiltonian. **Ψ** is the wave function that describes the system state from the coordinates of the positions of nuclei and electrons of the atoms in the molecule. E is the eigenvalue, and represents the total energy of the system.

2.2.1 Computational modeling of AChEIs

The theoretical studies of AChEIs were performed using the Gaussian03 program package (Frisch, 2004) in order to determine the best electronic and geometrical parameters of AChEIs molecules. Was have used DFT at the B3LYP hybrid functional level with 6-31+G(d,p) basis set. The geometries of the target drugs were first optimized using internal coordinates in order to minimize all geometric data.

AChEI structures were taken from Protein Data Bank (PDB). Specifically, the structure of AChEI complexed with AChE included the inhibitors tacrine (PDB code 1ACJ), galantamine (PDB code 1DX), donepezil (PDB code 1EVE), tacrine dimer (PDB code 2CKM), huperzine A (PDB code 1VOT) and rivastigmine (PDB code 1GQR). However, the three-dimensional structures of physostigmine, phenserine, metrifonate and dichlorvos were not found in a complex with AChE, and so the structures were modeled using the GaussView 4.1 program.

In order to analyze the activity of such AChEIs, we have computed the electronic and structural properties of those molecules: dipole, frontier molecular orbital energies Highest Occupied Molecular Orbital and Lowest Unoccupied Molecular Orbital (HOMO, HOMO-1, LUMO and LUMO+1), charge of heteroatoms, charge of most acid hydrogens, volume, distance between the most acid hydrogens – H-H, molecule size, LogP (partition coefficient/lipophilicity), LogS (solubility), number of H-bond acceptors, number of H-bond donors, number of aromatic rings, gap (HOMO-LUMO), and map of electrostatic potential (MEP). The charge used was from the ChelpG (i.e., the charge from electrostatic potentials using a grid-based method) population analysis.

2.3 Electronic structure of AChEI molecules

Table 1 shows the values for some calculated properties for the studied AChEIs. The values of the HOMO orbital energies are different when comparing the organophosphate DDVP and METRI (higher values) and other AChEIS. Thus, the values of the orbital HOMO-1 energy for these organophosphates are almost similar while the values of this energy for other AChEIS, structurally different, are between -6.04 and -6.75, indicating that this orbital may be important in process of interaction of these drugs with acetylcholinesterase.

Properties such as volume, size of the molecule, logP and dipole have different values for each AChEI. Although some of these molecules are of the same class, the calculated values for the same property changes sharply, as the volume between DDVP and METRI.

	DDVP	METRI	DIMTHA	THA	HUPE	GALA	E2020	PHYSO	RIVA	PHEN
Dipole (D)	3.04	2.93	2.09	3.51	6.08	1.67	2.48	1.08	2.53	1.85
Charge H$^+$ (ua)	0.070	0.438	0.260	0.339	0.365	0.371	0.136	0.333	0.103	0.281
Charge N (ua)	0	0	-0.320	-0.766	-1.013	-0.632	-0.484	-0.619	-0.378	-0.753
Charge O(ua)	-0.638	-0.639	0	0	-0.670	-0.685	-0.554	-0.659	-0.594	-0.505
HOMO (eV)	-6.86	-8.11	-5.90	-5.76	-5.90	-5.53	-5.95	-5.46	-5.79	-5.23
HOMO-1 eV)	-8.57	-8.30	-6.71	-6.56	-6.75	-6.04	-6.05	-6.08	-6.53	-6.43
LUMO (eV)	-0.54	-1.59	-1.47	-1.28	-1.30	-0.48	-1.53	-0.33	-0.43	0.52
LUMO+1(eV)	0.42	0.70	-0.58	-0.46	0.36	-0.14	-0.48	-0.07	0.21	0.51
GAP (eV)	6.32	6.52	4.47	4.49	4.60	5.53	4.41	5.12	5.36	5.75
H-H (Å)	1.796	2.307	1.998	1.683	1.625	2.373	2.342	2.324	2.460	2.250
Volume (Å³)	185	207	606	236	286	329	454	321	312	332
Size (Å)	7.849	7.068	19.386	9.516	9.051	10.290	12.281	12.927	11.242	14.808
logP	1.66	0.80	3.88	3.13	2.60	1.39	4.14	1.94	2.86	2.99
logS	-1.44	-1.77	-4.98	-3.16	-2.47	-2.23	-4.93	-2.44	-1.89	-4.2
PSA(Å²)	44.8	55.8	58.5	38.9	58.9	41.9	38.8	44.8	32.8	44.9
H-donnor	0	1	4	2	2	1	0	1	0	1
H-acceptor	4	4	4	2	4	4	4	5	4	5
Aromatic ring	4	0	4	2	1	2	2	2	1	1

H$^+$: most acid hydrogen of the molecule. H-H: distance between the most acid hydrogens.

Table 1. Electronic and structural data of AChEIs from B3LYP/6-31+G(d,p) level of calculation.

The distribution of frontier orbitals in the molecule is important to determine which atoms are most relevant for the interaction between the ligand and the receptor, during the activity of such drug with the biological target. Figure 3 shows the regions of the frontier orbitals of RIVA. The LUMO of RIVA is located in an unusual region for orbitals poor in electrons, since the benzyl ring is by definition the region of high electronic density. RIVA have the HOMO and LUMO in the opposite region of carbamate moiety.

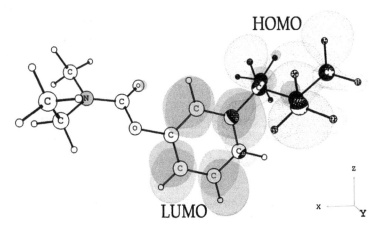

Fig. 3. HOMO and LUMO of RIVA calculated at B3LYP/6-31+G (d,p) level.

Figure 4 shows the MEP for all drugs studied. The maps indicate that all AChE have well-defined regions of high electron density (red) and low density (dark blue). These regions are likely points of interaction with AChE through electrostatic interactions like hydrogen bonding. Other electron density regions of AChE present mostly negative, indicating that these regions may have electrostatic interactions such as charge transfer, interactions with $\pi-\pi$ systems of aromatic residues in the active site of AChE. According to the literature (2003; de Paula et al., 2009; Nascimento, et al., 2011) RIVA interacts with the Ser200 residue of the catalytic triad through the carbamoyl moiety.

The regions of highest electronic density are located under the oxygen atoms in DDVP, METRI, HUPE, GALA, E2020, PHYSO, RIVA suggesting that these oxygens are receptors of hydrogens that can share in interactions, such as hydrogen bonding with the residues of the AChE catalytic triad or promoting covalent bonds, with the hydroxyl group of Ser200 (Camps, 2002). The rivastigmine (Figure 4h) and donepezil (Figure 4f) have more regions with low electronic density than other drugs, which is important for the interaction with the receptor site. The molecular volume is an important descriptor for drug design, and it is shown that some inhibitors are more linear which should improve the binding to two or more parts of the enzyme.

2.4 Mapping of the pharmacophoric profile of AChEIs via PCA

The mapping of pharmacophoric profile of AChEIs was conducted by means of PCA analysis, using of the data set obtained from calculations of the properties obtained for each molecule. For the analysis of principal components the geometry optimizations were performed at the level B3LYP/6-31 + G (d, p). The properties amount is shown in Table 1.

PCA was used to correlate 18 properties of 10 AChEI molecules to reduce the initial parameter set and determine the most relevant data for the acetylcholinesterase inhibition. PCA was conducted using the autoscaling method because the structural and electronic properties have different dimensions.

Since the structural and electronic properties have very different dimensions, autoscaling method is required. This normalization method consists in moving and stretching axes, centering the data on average and divides them by their standard deviation. Thus, each variable present zero mean and variance equal 1, giving the same significance for all variables (Jolliffe, 2002).

The first PCA showed that 83.3% on average of the information set formed by 10 AChEIs and 18 properties, obtained through the calculation performed in this work can be represented by four principals components (38.3% variance to PC1; 59.2% cumulative variance to PC2, 73.2% cumulative variance to PC3, 83.3% cumulative variance to PC4).

In studies of systems of many variables with PCA, it is desirable the lowest possible number of principal components that properly represent over 90% of cumulative variance. Thus, to increase the accuracy and determine the principal components we have used all combinations of 18 properties for the 10 AChEIs studied. Then the PCA was taken from the most relevant properties in relation to the total variance with the lower number of principal components.

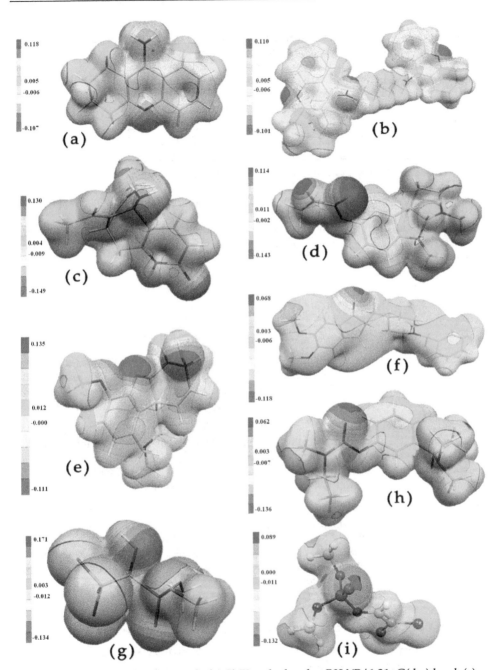

Fig. 4. Map of electrostatic potential of AChEIs calculated at B3LYP/6-31+G(d,p) level. (a) tacrine, (b) tacrine dimer, (c) huperzine A, (d) phenserine, (e) galantamine, (f) donepezil, (g) metrifonate, (h) rivastigmine, (i) dichlorvos.

In a second part, only 6 variables were used (shown below) with the 10 objects (AChEIs), these results in a 92% of total variance with information that characterizes these drugs in three principal components. Therefore, PCA (Figure 5) suggests that following properties are responsible for 92% of variance (63.5% for PC1 and 19.1% for PC2): volume, size of the molecule, distance between the two most acid hydrogens, energy of HOMO-1 frontier orbital, logP partition coefficient, and the number of aromatic rings (Table 2). In other words, these properties describes well the whole set of molecules data.

	PC1	PC2	PC3
B3LYP/6-31+G(d,p)	63.5 %	82.6%	91.7%

Table 2. Total cumulative variance (%) using 10 AChEIs and 6 properties

Equations 5, 6 and 7 show the scores of the PC1, PC2 and PC3. The PC1 is essentially the volume, size, partition coefficient- log P and number of aromatic rings of the drug (structural and electronic parameters), the PC2 represents the distance H-H (structural parameter) while the PC3 mainly represents the energy of the HOMO-1 orbital (electronic parameter).

$$PC1 = 0.48_{volume} + 0.46_{Size} + 0.13_{H-H} + 0.37_{HOMO-1} + 0.42_{logP} + 0.44_{Arom.} \quad (5)$$

$$PC2 = -0.01_{volume} + 0.02_{Size} + 0.88_{H-H} + 0.28_{HOMO-1} - 0.33_{logP} - 0.19_{Arom.} \quad (6)$$

$$PC3 = -0.38_{volume} - 0.40_{Size} - 0.16_{H-H} + 0.79_{HOMO-1} + 0.18_{logP} + 0.04_{Arom.} \quad (7)$$

Figure 5 shows that the molecules are grouped into some main groups in regard to PC1, one of these are the GALA, RIVA, PHEN and PHYSO which forms a cluster and these are the FDA approved drugs. HUPE, THA form other group with two potent inhibitors acetylcholinesterase. Therefore, volume, $\pi-\pi$ systems and drug size are important parameters that correlate the AChEIs.

On other hand PC2 is dominated by the HH distance (score: +0.88), which separates compounds according to the distance between the two hydrogens more acidic in two groups: one that found them have H-H values smaller than 2.0 Å, and other group of objects that have values greater than those cited.

Figure 6 depicts the scores of PC1 versus PC3. In this case there are two standards in PC3 DDVP / METRO and GALA / PHYSO / RIVA / PHEN, this is reasonable, since PC3 represent the orbital energy of HOMO-1 (+0.79) (Equation 7), these AChEIs have values close to the HOMO-1 and the volume, as shown in Table 1. Figures 5 and 6, Equations (5, 6 and 7) generated in the PC indicates that the electronic parameters - orbital energy of HOMO-1 , log P and number of aromatic systems - and the structural parameters - volume, size of the drug and H-H- are the most significant properties of AChE in this multivariate analysis.

3. Applications of PCA from pharmacophoric profile of the AChEIs

From the pharmacophore profile determined by means of PCA and electronic structure calculations, other classes of molecules can be considered to be classified as AChEI. Therefore, it is important to note that, when added other objects, the new plotted PCs should have accumulative variance equal or larger than the original PCA (Table 3). That is, the original data set is the reference.

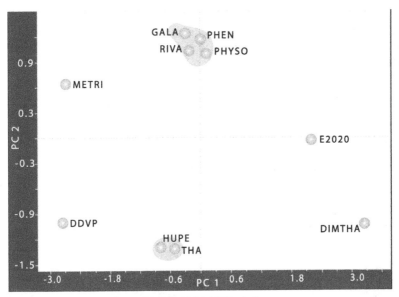

Fig. 5. PC1 versus PC2 scores from B3LYP/6-31+G(d,p) data.

	PC1	PC2	PC3
B3LYP/6-31+G(d,p)	71 %	86%	94%

Table 3. Total cumulative variance (%) using 10 AChEIs, 2 new molecules, 1 substrate and 6 properties.

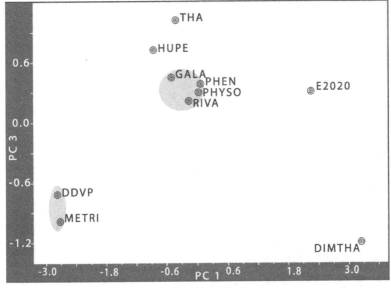

Fig. 6. PC1 versus PC3 scores from B3LYP/6-31+G(d,p) data.

The following are some applications explored. One possible application is the identification of dual biological activity of molecules. Another is the identification of activity against the function of acetylcholinesterase inhibitor, i.e., how behaves the natural substrate of the enzyme.

MOL1 and MOL2 with different biological activity from AChEIs were studied. We have also added another compound on the data set, acetylcholine (acetylcholinesterase substrate). Within these data from the properties obtained through electronic structure calculations, the new PCA was generated. These three objects were added to the 10 other objects of the first PCA which determined the pharmacophoric profile of AChEIs.

As shown in Figure 7, the natural substrate reorients the axis of the PCs and is completely opposed to the other (active) drugs. The acetylcholine is close to some of AChEIS by PC1, which have the properties of volume, size and number of aromatic rings with more high scores. Although acetylcholine is not an inhibitor it must have some properties correlated with the AChEIs. All molecules that are recognized by a protein have a certain set of properties that allows this recognition.

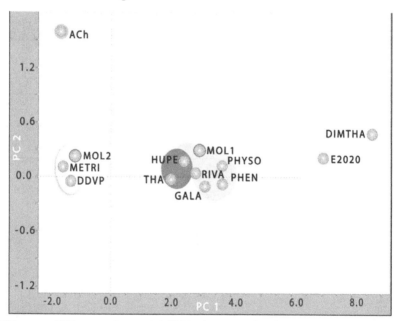

Fig. 7. PC1 versus PC2 scores from 10 AChEIs, 2 new molecules and ACh at B3LYP/6-31+G(d,p) level.

According to the PCA the two new drugs can be classified as AChEIs, although it is not possible to determine whether they have efficacy in the treatment of Alzheimer's disease. It is important to note that the effectiveness of a drug depends also on other factors that are difficult or much complex for modeling.

As shown in Figure 7, the natural substrate reorients the axis of the PCs and is completely opposed to the other (active) drugs. The acetylcholine is close to some of AChEIs by PC1, which have the properties of volume, size and number of aromatic rings with large scores. Although acetylcholine is not an inhibitor it must have some properties correlated with the AChEIs.

All molecules that are recognized by a protein have a certain set of properties that allows this recognition. As shown in Figure 8, in the case of acetylcholinesterase specifically, molecular modeling studies showed that during the molecular recognition of acetylcholine, the quaternary nitrogen of the substrate has a strong ionic interaction with the carboxyl group of the residue Asp72. In addition, there is formation of hydrogen bonds between the amine and ester group of the residue Asn65 and Tyr130 (Patrick, 2005).

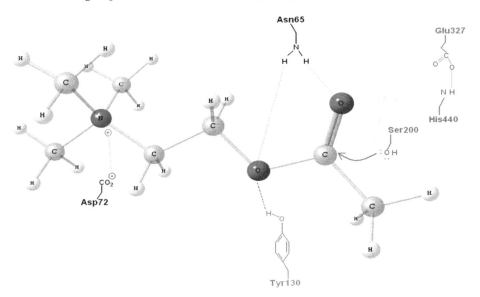

Fig. 8. PC1 Optimized structure of acetylcholine and main points of molecular recognition in acetylcholinesterase at B3LYP/6-31+G(d,p) level.

4. Conclusion

All studied AChE inhibitors are recognized as drugs that act effectively in the treatment of AD, even if some are not recommended due to unwanted side effects. From the computer modeling, applied by electronic structure calculations and PCA of known AChEIs with potential for the treatment of AD, it is possible to conclude that the properties relevant to the inhibitory activity are:

1. The HOMO-1 orbital is the most important property, characteristic of all drugs.
2. The study of the electrostatic potential map indicates that all AChEIs have well-defined regions of high and low electronic density, and these regions are the likely points of interaction with acetylcholinesterase, through electrostatic interactions such as hydrogen bonding. In some cases, such as dichlorvos, rivastigmine and physostigmine, it is possible to predict a bond with covalent character to the catalytic triad of the enzyme.

3. The H-H distance between 1.625-2.460Å, volume, and logP are also relevant for the pharmacophoric profile.
4. The electronic properties - orbital energy of the HOMO-1, log P and number of aromatic systems - and the structural parameters volume, size of the drug and H-H distance are the most significant properties in this study, the main components of the pharmacophore profile of AChEIs.
5. From the pharmacophore profile determined using PCA and electronic structure calculations, other classes of molecules can be considered to be classified as AChEI.

In addition to the structural and electronic properties mentioned above, an efficient inhibitor should have good bioavailability, cross the blood brain barrier with relative ease, to have high selectivity for acetylcholinesterase compared to butilcholinesterase; be able to form a complex reversible or pseudo-reversibly with AChE, be non-competitive compared to natural substrate. All these properties are jointly sharing in the profile of AChE pharmacophore studied in this work.

5. Acknowledgments

The authors are indebted to CNPq, INCTMN, FINEP, and CAPEs for financial support and Laboratório de Química Computacional/Universidade de Brasília (LQC/UnB) for computation resource.

6. References

Barak, D.; Ordentlich, A.; Kaplan, D.; Kronman, C.; Velan, B. & Shafferman, A. (2005). Lessons from functional analysis of AChE covalent and noncovalent inhibitors for design of AD therapeutic agents. *Chemico-Biological Interactions*, Vol.157, No.SI, (December 2005), PP 219-226, ISSN 0009-2797

Bartolucci, C.; Siotto, M.; Ghidini, E.; Amari, G.; Bolzoni, P. N.; Racchi, M.; Villetti, G.; Delcanale, M. & Lamba, D. (2006). Structural Determinants of Torpedo californica Acetylcholinesterase Inhibition by the Novel and Orally Active Carbamate Based Anti-Alzheimer Drug Ganstigmine (CHF-2819). *Journal of Medicinal Chemistry*, Vol.49, No.17, (August 2006), pp.5051-5058. ISSN 0022-2623

Camps, P. & Munoz-Torrero, D. (2002). Cholinergic Drugs in Pharmacotherapy of Alzheimer´s Disease. *Mini Reviews in Medicinal Chemistry*, Vol.2, No.1, (February 2002), pp.11-25, ISSN 1389-5575

Camps, P.; Formosa, X.; Munoz-Torrero, D.; Petrignet, J.; Badia, A. & Clos, M. V. (2005). Synthesis and pharmacological evaluation of huprine-tacrine heterodimers: Subnanomolar dual binding site acetylcholinesterase inhibitors. *Journal of Medicinal Chemistry*, Vol.48, No.6, (March 2005), pp. 1701-1704, ISSN 0022-2623

Carotti, A.; Altornare, C.; Savini, L.; Chlasserini, L.; Pellerano, C.; Mascia, M. P.; Maciocco, E.; Busonero, F.; Mameli, M.; Biggio, G. & Sanna, E. (2003). High affinity central benzodiazepine receptor ligands. Part 3: Insights into the pharmacophore and pattern recognition study of intrinsic activities of pyrazolo[4,3-c]quinolin-3-ones. Bioorganic & Medicinal Chemistry, Vol.11, No.23, (November 2003), pp. 5259-5272, ISSN 0968-0896

Costantino, H. R.; Leonard, A. K.; Brandt, G.; Johnson, P. H. & Quay, S. C. (2008). Intranasal administration of acetylcholinesterase inhibitors. *Bmc Neuroscience*, Vol.9, No.3, (December 2008), pp. 1-3, ISSN 1471-2202

Cramer, C. J. (November 22, 2004). *Essentials of Computational Chemistry* (2ⁿᵈ), Wiley, ISBN 047-0091-82-7, West Sussex

de Paula, A. A. N.; Martins, J. B. L.; dos Santos, M. L.; Nascente, L. D.; Romeiro, L. A. S.; Areas, T.; Vieira, K. S. T.; Gamboa, N. F.; Castro, N. G. & Gargano, R. (2009). New potential AChE inhibitor candidates. *European Journal of Medicinal Chemistry*, Vol.44, No.9, (September 2009), pp. 3754-3759, ISSN 0223-5234

de Paula, A. A. N.; Martins, J. B. L.; Gargano, R.; dos Santos, M. L. & Romeiro, L. A. S. (2007). Electronic structure calculations toward new potentially AChE inhibitors. *Chemical Physics Letters*, Vol.446,No.4-6, (October 2007), pp. 304-308, ISSN 0009-2614

Frisch, M. J., G.W. Trucks, H.B.S., G.E. Scuseria, M.A. Robb, J.R.C., Zakrzewski, V.G., Montgomery, Jr. J.A., Stratmann, R.E., Dapprich, J.C.B.S., Millam, J.M., Daniels, A.D., Strain, K.N.K.M.C., Farkas, O., Tomasi, J., Barone, V., Cammi, M.C.R., Mennucci, B., Pomelli, C., Adamo, C., Clifford, S., Petersson, J.O.G.A., Ayala, P.Y., Cui, Q., Morokuma, K., Rabuck, D.K.M.A.D., Raghavachari, K., Foresman, J.B., Ortiz, J.C.J.V., Stefanov, B.B., Liu, G., Liashenko, A., Komaromi, P.P.I., Gomperts, R., Martin, R.L., Fox D.J., Al-Laham, T.K.M.A., Peng, C.Y., Nanayakkara, A., Gonzalez, C., Challacombe, M., Gill, P.M.W., Johnson, B., Chen, W., Wong, M.W., Andres, J.L., Gonzalez, C., Head-Gordon, M., Replogle E.S. & Pople, J.A. (2003), Gaussian03, Gaussian Inc.,Pittsburgh, PA

Jolliffe, I. T. (October 1, 2002). *Principal Component Analysis* (2ⁿᵈ), Springers, ISBN 038-7954-42-2, New York

Lima, P.; Bonarini, A. & Mataric, M. (2004). *Application of Machine Learning*, InTech, ISBN 978-953-7619-34-3, Vienna, Austria

Nascimento, E. C. M. & Martins, J. B. L. (2011). Electronic structure and PCA analysis of covalent and non-covalent acetylcholinesterase inhibitors. *Journal of Molecular Modeling*, Vol.17, No.6, (June 2011), pp. 1371-1379, ISSN 1610-2940

Nascimento, E. C. M.; Martins, J. B. L.; dos Santos, M. L. & Gargano, R. (2008). Theoretical study of classical acetylcholinesterase inhibitors. *Chemical Physics Letters*, Vol.458,No.4-6, (June 2008), pp. 285-289, ISSN 0009-2614

Patrick, G. L. (2005) *An Introduction to Medicinal Chemistry* (3ʳᵈ), Oxford University Press, ISBN 019-9275-00-9, New York

Rocha, J. R.; Freitas, R. F. & Montanari, C. A. (2011). The GRID/CPCA approach in drug Discovery. *Expert Opinion on Drug Discovery*, Vol.5, No.4, (April 2011), pp. 333-346, ISSN 1746-0441

Sippl, W.; Contreras, J. M.; Parrot, I.; Rival, Y. M. & Wermuth, C. G. (2001). Structure-based 3D QSAR and design of novel acetylcholinesterase inhibitors. *Journal of Computer-Aided Molecular Design*, Vol.15, No.5, (May 2001) pp.395-410, ISSN 0920-654X

Steindl, T. M.; Crump, C. E.; Hayden, F. G. & Langer, T. J. (2005). Pharmacophore modeling, docking, and principal component analysis based clustering: Combined computer-assisted approaches to identify new inhibitors of the human rhinovirus coat protein. *Journal of Medicinal Chemistry* , Vol. 48, No.20, (October 2005), pp. 6250-6260, ISSN 0022-2623

Steindl, T. M.; Schuster, D.; Wolber, G.; Laggner, C. & Langer, T. J. (2006). High-throughput structure-based pharmacophore modelling as a basis for successful parallel virtual screening. *Journal of Computer-Aided Molecular Design*, Vol.20, No.12, (December 2006), pp. 703-715, ISSN 0920-654X

Sugimoto, H.; Ogura, H.; Arai, Y.; Iimura, Y. & Yamanishi, Y. (2002). Research and development of donepezil hydrochloride, a new type of acetylcholinesterase inhibitor. *Japanese Journal of Pharmacology*, Vol.89, No.1, (May 2002), pp. 7-20, ISSN 0021-5198

Tasso, S. M.; Bruno-Blanch, L. E.; Moon, S. C. & Estiú, G. L. (2000). Pharmacophore searching and QSAR analysis in the design of anticonvulsant drugs. Journal of Molecular Structure-Theochem, Vol. 504, No. , (June 2000), pp. 229-240, ISSN 0166-1280

Tezer, N. (2005). Ab initio molecular structure study of alkyl substitute analogues of Alzheimer drug phenserine: structure-activity relationships for acetyl- and butyrylcholinesterase inhibitory action. *Journal of Molecular Structure-Theochem*, Vol.714, No.2-3, (February 2005), pp. 133-136, ISSN 0166-1280

Ul-Haq, Z.; Khan, W.; Kalsoom, S. & Ansari, F. L. (2010). In silico modeling of the specific inhibitory potential of thiophene-2,3-dihydro-1,5-benzothiazepine against BChE in the formation of beta-amyloid plaques associated with Alzheimer's disease. *Theoretical Biology and Medical Modelling*, Vol.7, No. 22, (June 2010), pp. 2-26, ISSN 1742-4682

Wermuth, C. G.; Ganellin, C. R.; Lindberg, P. & Mitscher, L. A. (1998). Glossary of terms used in medicinal chemistry (IUPAC Recommendations 1998). *Pure and Applied Chemistry*, Vol.70, No.5, (May 1998), pp. 1129–1143, ISSN: 0033-4545

6

Application of PCA in Taxonomy Research – Thrips (Insecta, Thysanoptera) as a Model Group

Halina Kucharczyk[1], Marek Kucharczyk[2],
Kinga Stanisławek[1] and Peter Fedor[3]
[1]Department of Zoology, Maria Curie-Skłodowska University, Lublin,
[2]Department of Nature Conservation, Maria Curie-Skłodowska University, Lublin,
[3]Department of Ecosozology, Faculty of Natural Sciences, Comenius University, Bratislava,
[1,2]Poland
[3]Slovakia

1. Introduction

Phenetic taxonomy is based on analysis of many unweighted characters. The number of variables that can be analyzed for a plant or animal species is so high that it is necessary to use a mathematical tool for grouping them into units corresponding to taxa.

Principal Component Analysis enables researchers to reduce the number of possible groupings. It is significant due to the occurrence of some redundancy in variables. In this case, redundancy means that some of the variables are correlated with one another, because they are measuring the same construct.

Principal Component Analysis replaces many original characters with only a few most significant principal components (PCs) which represent combinations of closely correlated original characters.

Principal Component Analysis was first described by Pearson (1901); in the 1930's Hotteling (1933) prepared a fully functional method that generates a set of orthogonal axes, placed in decreasing order and determining the main directions of variability of samples.

In analysis of principal components, eigenvalues represent the relative participation of each principal component in presenting the general variability of sampled material. The numerical value of a given eigenvalue is a direct indicator of the weight of a particular component in the general characteristics of the variability of a set of data. In practice, the distribution of the elements of the analyzed set in the space of the first three or four components allows one to present almost the complete diversity of the set.

Eigenvectors are sets of numbers which show the weights of individual characters for each principal component. Like the correlation coefficient, eigenvalues are scaled from -1 to +1. The higher the value, the more closely a given trait is connected with a component. On the basis of eigenvectors it is then possible to interpret principle components, e.g. determine which character (or characters) are the most representative.

Before principal components are determined, the data can be processed in many ways. It is necessary if variables (features) are expressed in different units or the range of their variability is different. The two methods used are centering and standardization. In the former, the variable axes are moved so that the beginning of one axis is in the centre of inertia of the axis. Standardization of data involves changing the values of characters in such a way that their mean is 0 and standard deviation is 1.

If analysis covers calculating the covariance matrix, the variables are centred; if, on the other hand, the correlation coefficient matrix needs to be calculated, the variables are centred and standardized at the same time.

The Principal Component Analysis (PCA) is widely used in taxonomic research of plants and animals (Apuan et al., 2010; Chiapella, 2000; Kucharczyk & Kucharczyk 2009; Lilburn & Garrity, 2004; Sahuquillo et al., 1997; Wolff et al., 1997; Véla et al., 2007). Below we present application of PCA in taxonomy of the order Thysanoptera (Insecta). The numerical calculations were made with the Multi-Variate Statistical Package (MVSP) (Kovach, 2005) and Statistica PL, version 6 (StatSoft Inc. 1984-2001).

1.1 Characteristic of the order Thysanoptera

Up to now almost 6000 species of the order Thysanoptera have been described worldwide but many other are added to this list every year. The asymmetric, with only one left mandibule, mouth cone is a synapomorphic character state which differs thrips from the other insect orders (Mound et al., 1980; Mound & Morris, 2004, 2007). In currently accepted systematic position the order Thysanoptera is divided into two suborders: Terebrantia and Tubulifera. The former consists of eight families which comprise very small and tiny insects, most of them reaching 1-3 mm in length. They are mostly herbivorous insects feeding both on dicotyledonous and monocotyledonous plants, only a small part of them are facultative or obligate predators (fam. Aeolothripidae). The family Thripidae with nearly 2,500 species is the largest within Terebrantia but the relationships within this group is not clear (Mound & Morris, 2007). Only one tubuliferan family – Phlaeothripidae with almost 3,500 known species includes larger thrips, the biggest, mostly tropical taxa, reach up to 15 mm. Excepting herbivores most of them live on dead wood or in leaf litter and feed on fungi. Some phytophagous species are regarded as pests, feeding and breeding on different parts of plants they cause deformations of leaves, flowers and fruits, and in the final result stop their development. A limited number of terebrantians may transmit fungi, bacteria and viruses, which may infect the host plants reducing the quality of yields and their market value (Lewis, 1997; Tommasini & Maini, 1995).

Originally taxonomy referred to the description and naming of the organisms (alpha taxonomy). Currently it is the science based on different fields of knowledge and that uses various tools to classify the organisms and determine the relationships amongst them. There are two systems of the Thysanoptera order classification: phenetic based mainly on the morphological characteristics of adults' specimens and phylogenetic one based on the evolutionary relationships (Mound, 2010). The former one is more practical and widely used in construction of the identification keys. The latter based on the molecular and genetic data is not satisfactory enough because of insufficient data. Therefore in practice, morphology and other biological aspects, e.g. observations of developmental stages and relations with host plants provide more data and may be useful in comprehension of the relationships amongst taxa (Crespi et al., 2004; Mound & Morris, 2007).

Some disagreements exist concerning classification system of Thysanoptera on different taxonomic levels. Because of the great differences in body structures between Terebrantia and Tubulifera Bhatti (1988) proposed to raise them to the order rank in new superorder Thysanopteroidea composed of 40 families. In the next works this author divided Terebrantia into 28 families (Bhatti, 2005, 2006). However, this classification is not accepted by most of thysanopterologists now. On the other hand the current state of Terebrantia with eight families has not been taken under consideration by zur Strassen (2003) in his latest key. In contrast to the currently accepted division zur Strassen classified the species of the genera *Melanthrips* and *Ankothrips* into the family Aeolothripidae. Many revisions at the genus level took place in the past, e.g. changes within the genera: *Thrips* Linnaeus 1758 and *Taeniothrips* Amyot & Serville 1843, *Mycterothrips* Trybom 1910 and *Taeniothrips*, *Anaphothrips* Uzel 1895 and *Rubiothrips* Schliephake 1975 etc. (Strassen zur 2003).

The correct identification of specimens to the species level is a basis for further taxonomic study. Often there are many problems with recognition of adults, as well as immature stages because of diversity of variation within and between species, particularly in the species rich genera, e.g. *Thrips* (Terebrantia) and *Haplothrips* Amyot & Serville 1843 (Tubulifera). These genera include mainly the Holarctic species feeding and breeding on dicotyledonous plants, though occasionally they are graminicolous. Very numerous species representing both of them may suggest that these genera are relatively young in evolutionary history. Morphologically, many species are very similar and are treated by some researchers as the same species in two forms, e.g. *Thrips atratus* Haliday 1836 and *Thrips montanus* Priesner 1921, *Thrips sambuci* Heeger 1854 and *Thrips fuscipennis* Haliday 1836, *Thrips fuscipennis* and *Thrips menyanthidis* Bagnall 1923, *Haplothrips leucanthemi* (Schrank 1781) and *Haplothips niger* (Osborn 1883), *Haplothrips tritici* (Kurdjumov 1912) and *Haplothrips cerealis* Priesner 1939 (Mound & Minaei, 2007, 2010; Strassen zur, 2003).

There are even more problems with recognizing larval stages. The larvae are less mobile therefore have a stronger relationship with their host plants. Morphological dissimilarity amongst larvae of different species are often larger than amongst their adults (Kucharczyk, 2010). So, the detailed study both on adults and immature stages with application of the statistical tools, within PCA method, allows to solve several problems in the taxonomy of thrips and at least partially explain their phylogenetic relationships.

2. Application of Principal Component Analysis method in taxa recognizing

2.1 Identification of the second larval instar of the *Haplothrips* genus species

The Principal Component Analysis method (PCA) may be useful in selecting from among the great number of morphometric characters, especially those that have some taxonomical value. Such a necessity is occuring within genera which species are very uniform in morphological structure and there are weak qualitative characters differentiating them. The only potential differences are related to the measurements of some body parts. The genus

Haplothrips (Thysanoptera, Phlaeothripidae) is a good example for the discussed case. To this genus belong species that are very similar in body structure and therefore they are difficult to identify.

The *Haplothrips* genus is one of the most numerous in species of the Phlaeothripidae family (about 230 species). It is distributed worldwide but mainly in northern hemisphere, almost 70% known hitherto species of this genus have been noted in Holarctic. Most of them are phytophagous, feeding and breeding in flowers of dicotyledons, mostly of *Asteraceae* family, only a few taxa are connected with monocotyledons (Mound & Minaei, 2007; Pitkin, 1976; Zawirska & Wałkowski, 2000). Because of the fact that the research on larval morphology is very scarce (especially on species belonging to Phlaeothripidae) the morphological analysis of the second instar larvae of *Haplothrips* species have been undertaken.

The measurements were made for 165 specimens of larvae belonging to 11 species (each species was represented by 15 individuals) collected from various habitats and regions of Poland. Among examined species were: *Haplothrips aculeatus* (Fabricius 1803) (inhabitant of various *Poaceae* plants, the ubiquistic species); *H. tritici* (Kurdjumov 1912) (also *Poaceae*-related species, but mainly with *Triticum vulgare*; occurring in south-eastern part of Europe and Asia Minor); *H. setiger* Priesner 1921 (the polyphagous species living on flowers of *Asteraceae* for example *Senecio fuchsii, Crepis* spp., *Matricaria* spp., *Achillea* spp., *Anthemis arvensis*; having Palaearctic range of occurrence); *H. subtilissimus* (Haliday 1852) (present on leaves of bushes and trees such as *Quercus* spp. and *Fagus* spp.; considered as predatory species on mites, lepidopteran eggs and larval stages of *Coccidae* and *Aleyrodidae*; noted from Europe). The rest of species are regarded as mono- or oligophags: *H. arenarius* Priesner 1920 (especially connected with *Helichrysum arenarium*; with Palaearctic range); *H. dianthinus* Priesner 1924 (on *Dianthus carthusianorum*; noted from central and south Europe); *H. jasionis* Priesner 1950 (on *Jasione montana*; known from Europe only); *H. leucanthemi* (Schrank 1781) (on *Leucanthemum* spp.; reaching Holarctic, Oriental and Australian regions); *H. setigeriformis* Fabian 1938 (on *Potentilla argentea*, known from Europe and Near East); *H. statices* (Haliday 1836) (on *Armeria elongata*; occurring in Europe) and *H. angusticornis* Priesner 1921) (connected with *Asteraceae* but mainly with *Achillea millefolium*; noted from Palaearctic).

Each individual of larva have been measured in respect of 72 potentially important features, most of them concerned lengths of selected both dorsal (d) and ventral (v) setae on all parts of body (h – head, pro - pronotum, mes – mesonotum, meta – metanotum), distances between setae, measurements of apical abdominal segments (8-11) and antennal segments (ant, III-VII). Finally, the data matrix consisted of 11880 measurements was constructed (comprised of 72 character states of 165 individuals belonging to 11 species) (Fig. 1).

The specimens were ordinated along first two PCA axes (transformed data). The results of PCA showed that the cumulative variance of the two principal components reached 62.4%: Axis 1 - 38.5%, Axis 2 - 23.9%. Figure 2 shows two groups of specimens belonging to different species which are clearly isolated (*H. aculeatus* and *H. subtilissimus*). The most discriminative features with the highest eigenvalues are the lengths of abdominal setae (9-d2, 9-v2, 11-d1, 11-v2). On the mentioned PCA graph the most numerous group of specimens, belonging to the nine other and listed above species, is creating compact cloud. To choose more selective character states discriminating the rest of examined species we can remove the measurements of the two first separated species (*H. subtilissimus* and *H. aculeatus*) from the primary data matrix.

	ant-l	ant-dist	I-l	I-w	I-lAw	II-l	II-w	II-lAw	III-l	III-w	III-lAw	IV-l	IV-w	IV-lAw	V-l	V-w	V-lAw	VI-l	VI-w	VI-lAw	VII-l	VII-w	VII-lAw	t-3	t-4-int	t-4-ex	t-5	t-6
acul1	230,0	25,0	20,0	32,5	0,8	30,0	21,3	1,4	50,0	22,5	2,2	47,5	25,0	1,9	32,5	22,5	1,4	25,0	17,5	1,4	22,5	8,8	2,8	10,0	12,5	5,0	10,0	13,8
acul2	220,0	22,5	20,0	30,0	0,7	30,0	21,3	1,4	47,5	22,5	2,1	47,5	23,8	2,0	32,5	22,5	1,5	25,0	16,3	1,5	22,5	8,8	2,8	10,0	13,8	5,0	11,3	17,5
acul3	230,0	22,5	20,0	32,5	0,8	30,0	20,0	1,5	50,0	22,5	2,2	47,5	26,0	1,9	35,0	22,5	1,8	25,0	16,3	1,5	22,5	7,5	3,0	10,0	12,5	5,0	12,5	15,0
acul4	220,0	22,5	17,5	27,5	0,8	27,5	18,8	1,5	45,0	21,3	2,1	42,5	22,5	1,9	30,0	21,3	1,4	22,5	15,0	1,5	20,0	7,5	2,7	7,5	12,5	5,0	10,0	12,5
acul5	225,0	22,5	22,5	28,8	0,8	30,0	21,3	1,4	50,0	21,3	2,4	45,0	22,5	2,0	32,5	20,0	1,8	25,0	15,0	1,7	22,5	7,5	3,0	10,0	12,5	7,5	10,0	15,0
acul6	205,0	22,5	20,0	30,0	0,7	27,5	20,0	1,4	45,0	20,0	2,3	40,0	22,5	1,8	27,5	20,0	1,4	20,0	16,3	1,2	20,0	8,8	2,3	10,0	12,5	6,3	10,0	12,5
acul7	225,0	22,5	25,0	32,5	0,8	30,0	21,3	1,4	50,0	22,5	2,2	45,0	26,0	1,8	32,5	22,5	1,4	25,0	16,3	1,5	22,5	8,8	2,8	10,0	12,5	5,0	10,0	12,5
acul8	235,0	25,0	30,0	30,0	1,0	30,0	22,5	1,3	50,0	22,5	2,2	47,5	23,8	2,0	35,0	22,5	1,8	25,0	16,3	1,5	25,0	8,8	2,9	8,8	12,5	5,0	10,0	12,5
acul9	225,0	22,5	25,0	30,0	0,8	30,0	20,0	1,5	52,5	22,5	2,3	45,0	25,0	1,8	35,0	22,5	1,8	25,0	16,3	1,7	22,5	7,5	3,0	7,5	12,5	3,8	8,8	13,8
acul10	200,0	22,5	22,5	27,5	0,8	27,5	20,0	1,4	42,5	20,0	2,1	42,5	21,3	2,0	30,0	18,8	1,8	22,5	15,0	1,5	17,5	7,5	2,3	8,8	12,5	3,8	10,0	13,8
acul11	220,0	22,5	25,0	30,0	0,8	30,0	20,0	1,5	45,0	21,3	1,9	37,5	22,5	1,7	25,0	15,0	1,7	22,5	15,0	1,5	20,0	7,5	3,0	10,0	12,5	5,0	11,3	15,0
acul12	215,0	22,5	25,0	30,0	0,8	30,0	20,0	1,5	47,5	22,5	2,1	42,5	26,0	1,7	32,5	22,5	1,4	25,0	15,0	1,7	20,0	7,5	2,7	7,5	12,5	5,0	11,3	13,8
acul13	215,0	22,5	20,0	30,0	0,7	30,0	21,3	1,4	47,5	22,5	2,1	42,5	25,0	1,7	30,0	22,5	1,3	22,5	15,0	1,5	20,0	8,8	2,3	7,5	12,5	5,0	10,0	15,0
acul14	215,0	22,5	20,0	30,0	0,7	27,5	20,0	1,4	47,5	21,3	2,2	42,5	22,5	1,9	32,5	20,0	1,6	22,5	15,0	1,5	20,0	8,8	2,3	10,0	12,5	5,0	11,3	12,5
acul15	210,0	25,0	17,5	30,0	0,6	27,5	20,0	1,4	46,0	20,0	2,3	42,5	22,5	1,9	32,5	20,0	1,6	22,5	15,0	1,5	20,0	9,8	2,3	7,5	12,5	5,0	11,3	12,5
aren1	205,0	10,0	20,0	27,5	0,7	27,5	21,3	1,3	35,0	21,3	1,8	37,5	22,5	1,7	35,0	20,0	1,8	27,5	15,0	1,8	20,0	7,5	2,7	5,0	8,8	3,8	6,3	6,3
aren2	205,0	15,0	20,0	32,5	0,6	30,0	22,5	1,3	37,5	22,5	1,7	37,5	22,5	1,7	35,0	20,0	1,8	27,5	15,0	1,8	22,5	7,5	3,0	6,3	10,0	2,5	7,5	7,5
aren3	210,0	15,0	20,0	32,5	0,6	32,5	22,5	1,4	37,5	22,5	1,7	40,0	22,5	1,8	35,0	20,0	1,8	27,5	15,0	1,8	25,0	7,5	3,0	6,3	9,9	5,0	8,8	10,0
aren4	205,0	10,0	20,0	27,5	0,7	30,0	21,3	1,4	40,0	21,3	1,9	37,5	22,5	1,7	32,5	20,0	1,8	27,5	15,0	1,8	25,0	7,5	3,0	5,0	10,0	3,8	7,5	7,5
aren5	200,0	10,0	20,0	27,5	0,7	30,0	20,0	1,5	36,0	22,5	1,6	37,5	22,5	1,7	32,5	22,5	1,4	27,5	15,0	1,7	22,5	7,5	2,8	6,3	8,8	3,8	7,5	8,8
aren6	211,0	12,4	21,9	32,3	0,7	30,8	22,1	1,4	36,8	23,9	1,5	38,3	24,5	1,6	33,8	21,5	1,6	28,0	16,7	1,7	23,9	9,8	2,4	5,0	8,7	4,0	7,0	8,3
aren7	227,7	15,0	23,9	32,2	0,7	33,0	23,5	1,4	42,7	24,0	1,8	43,2	24,6	1,8	37,5	22,4	1,7	29,8	16,7	1,8	24,9	9,5	2,6	5,9	10,7	5,3	9,0	9,2
aren8	213,8	14,0	20,4	29,3	0,7	30,2	21,8	1,4	39,0	22,7	1,8	40,0	24,5	1,6	36,2	21,1	1,7	29,9	15,8	1,9	23,3	8,5	2,7	5,8	8,1	3,3	6,5	7,3
aren9	199,5	9,8	21,1	28,6	0,7	29,5	20,6	1,4	34,9	21,5	1,8	35,3	22,2	1,6	32,8	18,8	1,8	27,3	14,7	1,9	22,7	9,0	2,5	6,1	7,8	3,7	6,8	6,9
aren10	207,9	15,0	20,1	30,9	0,7	31,0	21,9	1,4	32,9	24,0	1,4	40,8	25,0	1,6	35,0	22,5	1,8	28,9	15,7	1,8	23,9	8,9	2,7	5,7	9,9	4,5	7,4	8,7
aren11	197,8	9,8	19,9	29,7	0,7	26,9	20,6	1,3	35,0	21,9	1,6	36,2	22,4	1,6	31,1	19,9	1,6	28,3	15,8	1,8	23,2	9,2	2,5	5,3	7,4	4,8	6,0	7,5
aren12	205,8	13,5	19,3	29,7	0,7	30,0	21,8	1,4	34,8	24,0	1,5	36,8	24,3	1,5	33,2	20,6	1,8	29,5	16,0	1,8	23,0	8,8	2,6	6,0	9,8	4,8	8,2	8,9
aren13	204,6	12,0	18,6	28,9	0,6	29,2	21,0	1,4	36,6	21,1	1,7	37,0	22,9	1,6	31,9	20,3	1,8	28,0	15,5	1,8	25,3	8,8	2,9	5,0	8,5	4,6	7,0	6,9
aren14	201,5	9,8	19,0	28,4	0,7	29,2	19,8	1,5	36,1	20,2	1,8	38,7	21,8	1,8	32,4	19,2	1,7	27,0	14,0	1,9	21,9	8,9	2,5	6,3	9,2	4,0	7,4	9,2
aren15	214,2	13,3	21,3	30,9	0,7	29,8	22,3	1,3	37,0	22,9	1,6	38,7	24,7	1,6	34,6	20,8	1,7	28,5	16,2	1,8	26,0	8,3	3,1	4,6	8,3	3,6	6,2	6,0
dianth1	265,0	17,5	22,5	35,0	0,8	35,0	22,5	1,5	57,5	25,0	2,3	50,0	25,0	2,0	46,0	22,5	2,0	32,5	17,5	1,9	22,5	10,0	2,3	7,5	12,5	2,5	10,0	10,0
dianth2	265,0	25,0	22,5	35,0	0,8	37,5	22,5	1,7	57,5	25,0	2,3	50,0	27,5	1,8	45,0	25,0	1,8	32,5	17,5	1,9	22,5	10,0	2,3	7,5	12,5	2,5	10,0	10,0
dianth3	272,5	17,5	25,0	35,0	0,7	37,5	22,5	1,7	60,0	25,0	2,4	52,5	25,0	2,1	42,5	22,5	1,9	35,0	17,5	2,0	22,5	8,8	2,6	7,5	12,5	2,5	7,5	10,0
dianth4	270,0	17,5	25,0	37,5	0,7	37,5	22,5	1,7	60,0	25,0	2,4	52,5	25,0	2,1	46,0	22,5	2,0	35,0	17,5	2,0	22,5	8,8	2,5	7,5	12,5	2,5	10,0	10,0
dianth5	275,0	20,0	25,0	35,0	0,7	35,0	23,8	1,5	57,5	25,0	2,2	46,0	23,8	1,9	32,5	20,0	2,4	35,0	17,5	2,0	22,5	10,0	2,0	7,5	12,5	3,8	8,8	10,0
dianth6	260,0	22,5	25,0	33,8	0,7	35,0	23,8	1,5	62,5	25,3	2,4	52,5	27,5	1,9	42,5	23,8	1,8	35,0	17,5	1,9	20,0	10,0	2,0	7,5	12,5	3,8	7,5	10,0
dianth7	270,0	20,0	22,5	35,0	0,8	35,0	25,0	1,4	60,0	25,0	2,4	52,5	26,3	2,0	42,5	23,8	1,8	35,0	17,5	2,0	22,5	10,0	2,3	7,5	12,5	3,8	8,8	10,0
dianth8	270,0	20,0	22,5	40,0	0,9	37,5	23,8	1,6	62,5	25,0	2,5	52,5	26,3	2,1	43,8	22,5	1,9	33,8	17,5	1,9	22,5	10,0	2,3	7,5	12,5	3,8	10,0	10,0
dianth9	265,0	22,5	22,5	37,5	0,8	37,5	22,5	1,7	60,0	25,0	2,4	50,0	25,0	2,0	42,5	22,5	1,9	32,5	17,5	1,9	22,5	10,0	2,3	7,5	12,5	3,8	10,0	7,5

Fig. 1. A piece of data matrix comprising (in the whole) 72 metric characters of the second larval instar of eleven *Haplothrips* species

The result of the next PCA is separation of the next group of individuals belonging to *H. dianthinus* species (Axis 1 – 40.5%, Axis 2 – 18.1%; Fig. 3). The most discriminative features are short setae 9-v1 and 11-d1. There are some additional selective features (e.g. 8-v2, h-s2) but they are less significant. After elimination of *H. dianthinus* data from the matrix next two species are emerging: *H. arenarius* and *H. angusticornis* (Fig. 4) and three main characters are discriminated (setae: 8-v2, 9-d2, 9-v2). Now we can establish the value-range of selective features for examined species both with the help of data matrix and PCA graph.

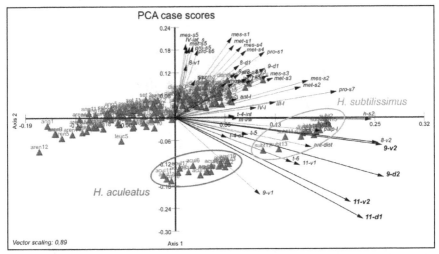

Fig. 2. PCA – scatter diagram of studied *Haplothrips* spp. specimens as OTUs along PC1 and PC2 based on 72 quantitative features (names of species and characters abbreviated)

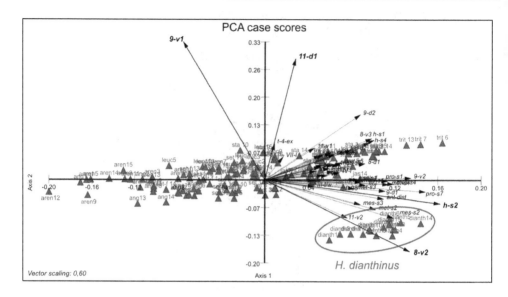

Fig. 3. PCA – scatter diagram of studied *Haplothrips* spp. specimens as OTUs along PC1 and PC2 based on 72 quantitative features (without specimens of *H. aculeatus* and *H. subtilissimus*)

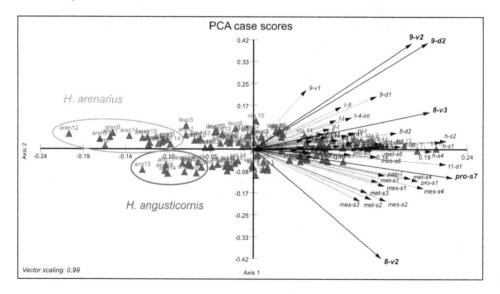

Fig. 4. PCA – scatter diagram of studied *Haplothrips* spp. specimens as OTUs along PC1 and PC2 based on 72 quantitative features (without specimens of *H. aculeatus*, *H. subtilissimus* and *H. dianthinus*)

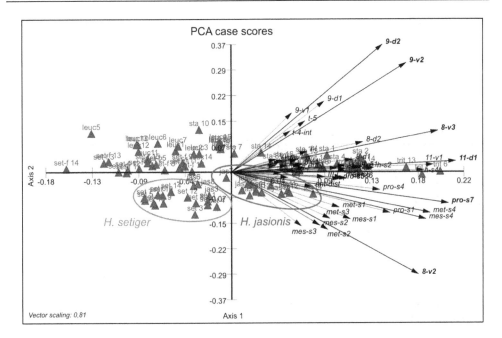

Fig. 5. PCA – scatter diagram of studied *Haplothrips* spp. specimens as OTUs along PC1 and PC2 based on 72 quantitative features (without specimens of *H. aculeatus*, *H. subtilissimus*, *H. dianthinus*, *H. arenarius* and *H. angusticornis*)

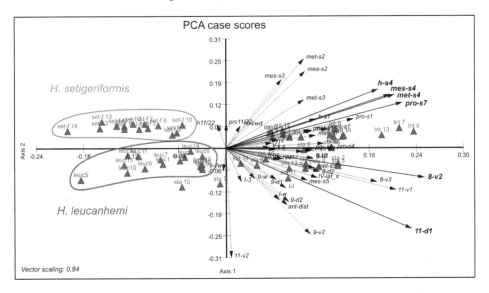

Fig. 6. PCA – scatter diagram of studied *H. setigeriformis*, *H. leucanthemi*, *H. tritici* and *H. statices* specimens as OTUs along PC1 and PC2 based on 72 quantitative features

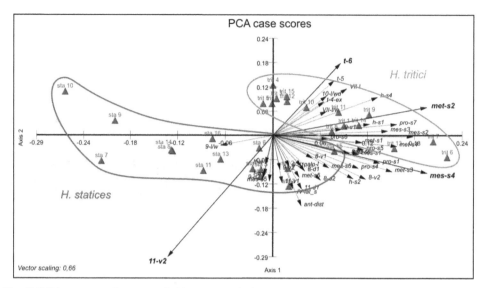

Fig. 7. PCA – scatter diagram of H. tritici and H.statices specimens as OTUs along PC1 and PC2 based on 72 quantitative features

Next steps of the analysis are similar to the previous ones (Figs. 5-7). During the analysis, it may turn out, that some specimens have ambiguous position (like "jas3, jas6, jas8" Fig. 5) and this may be a hint to check the data of the doubtful ones. At the end of the elimination, two species which are the most resembling (H. tritici and H. statices) remain. The selected distinguishing features can be used to construct an identification key. Additional benefit of presented method is elimination of characters which have no taxonomical value (in this case, out of 72 tested features there are 22 unimportant states). That would speed up further measurements of more specimens and improve the precision of determined features range.

2.2 Identification of the second larval instar of the *Thrips* genus species

The study on the morphology of the second larval instar of Central European *Thrips* species is another example of using PCA method in taxonomic research (Kucharczyk, 2010). In contrast to the previously discussed larvae of the genus *Haplothrips*, larvae of 34 researched species of the genus *Thrips* may be recognized mainly on the basis of qualitative characters. The data matrix covered 26 multistate discontinuous characters, amongst them the most important related to the sclerotisation and sculpture of integument, the structure of spiracles and antennae (Tab. 1). This analysis was conducted in two steps. At the beginning all researched specimens were treated as operational taxonomic units (OTUs) which have been characterized by 26 variables. On the graph OTUs form the clouds which are corresponding to the studied species. During the second step not specimens but species were treated as OTUs. In these cases the PCA was applied as a method for ordination and reducing the number of variables, the characters which had the highest loadings to PC1 and PC2 were extracted and they were regarded as the most important and useful in distinguishing the studied taxa (Fig. 8). Finally, these selected features have been used in constructing the identification key to second larval instar of studied *Thrips* species.

The characteristics tested in PCA method have been also used as variables in Claster Analysis (CA). The results in the two-dimension ordination of PCA were consistent with the results in the hierarchical clustering analysis (Fig. 9). The results of numerical analysis sheds some new light on the relationships amongst studied species. Moreover, the received dendrogram showed the similarity within studied *Thrips* species and allowed to propose the ancestral (plesiomorphic) and advanced (apomorphic) characteristics of immature stage which have not been studied hitherto.

Feature	Abbrev.	Feature	Abbrev.
head sclerotization	HPLATE	sculpture of tergite VIII	T8SCP
pronotum sclerotization	PRPLATE	tergite IX - sculpture anterior to and between campaniform sensilla (Cs)	T9SCP
mesonotum sclerotization	MSPLATE	tergite IX - sculpture between setae level and Cs	T9DSCP
metanotum sclerotization	MTPLATE	tergite IX - sculpture, teeth between D1 setae	T9D1TH
tergite IX sclerotization	T9SCL	sculpture of tergite X	T10SCP
tergite X sclerotization	T10SCL	microtrichial comb at base of hind legs	COXAT
sternite XI sclerotization	S11SCL	spiracles on abdominal segment II	AT2SP
posteromarginal comb on tergite IX	T9COMB	number of facets in spiracle on abdominal segment II	NOOFF
length of teeth on posteromarginal comb	LCOMB	width of spiracle on abdominal segment II	WOFSP
sculpture of mesonotum	MSSCP	microtrichia on antennal segment III	ANS3MT
sculpture of metanotum	MTSCP	length of trichome on antennal segment VI	ANS6TR
sculpture of tergites III-VII	T3-7SC	shape of abdominal tergal setae D1, D2	TSETA
sculpture of sternites III-VII	S3-7SC	furca on metasternum	FURCA

Table 1. Features of second larval instar of *Thrips* species used in numerical analyses (after Kucharczyk, 2010)

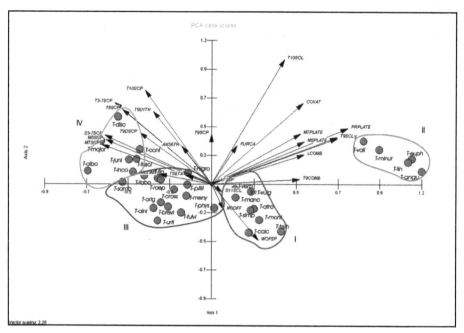

Fig. 8. PCA – scatter diagram of *Thrips* genus species as OTUs along PC1 and PC2 based on 26 qualitative features (names of species abbreviated, abbreviations of characters as in Table 1)

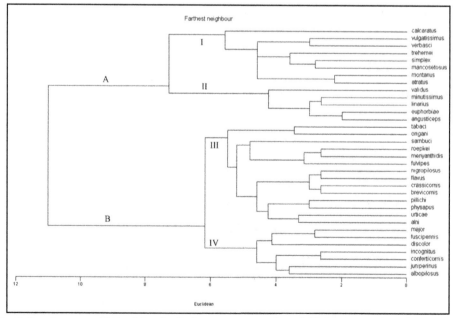

Fig. 9. Dendrogram (CA) of the similarity amongst analyzed *Thrips* species based on the morphological characters of the second larval instar (after Kucharczyk, 2010)

2.3 Identification of *T. atratus* Haliday 1836 and *T. montanus* Priesner 1920 species

For the first time Kucharczyk & Kucharczyk (2009) have used PCA method in the study on two similar *Thrips* species: *T. atratus* and *T. montanus*. Since the description, both of them were re-classified to *Thrips* Linnaeus 1758, *Taeniothrips* Amayot et Serville 1843 or *Similothrips* Schliephake 1972 genera (Priesner, 1964; Schliephake, Klimt, 1979; Schliephake, 2001). Zur Strassen (2003) replaced the latter species by *T. atratus* as its mountainous form.

During the study on the *Thrips* genus larvae the morphological differences between second larval instar of *T. atratus* and *T. montanus* were observed (Kucharczyk, 2010). These species also showed different food preferences. PCA method was applied for distinguishing the most important morphological, measurable features (8 for females and 12 for males) which may be useful in recognizing these species.

On the graphs prepared for females and males separately the specimens of both species were ordered along the first two principal components, the lengths of vectors were correlated with the features significant in recognizing the studied taxa (Figs. 10, 11, Tab. 2). Two principal components sequentially accounted for cumulatively 76.0% for females (62.7% and 13.3% for PC1 and PC2 respectively) and 63.6% for males (41.4% and 22.0% for PC1 and PC2 respectively). The quantitative characters together with the qualitative ones both of adults and larvae and the host preferences: polyphagy for *T. atratus* and monophagy (on *Rhinanthus* spp.) for *T. montanus* allowed to consider them as valid taxa.

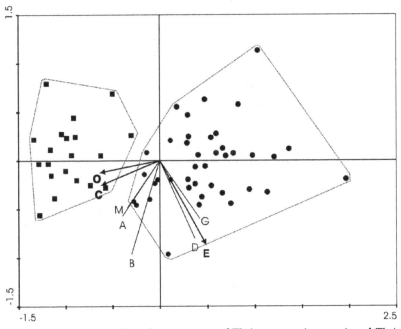

Fig. 10. PCA scatter diagram of female specimens of *Thrips atratus* (squares) and *Thrips montanus* (circles) as OTUs along PC1 and PC2 based on 8 quantitative characters (symbols as in Table 2) (after Kucharczyk & Kucharczyk, 2009)

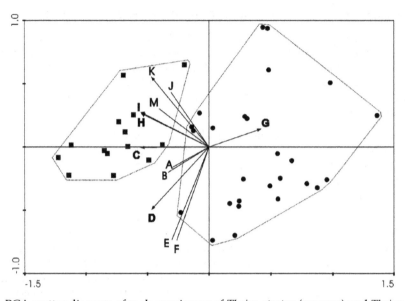

Fig. 11. PCA scatter diagram of male specimens of *Thrips atratus* (squares) and *Thrips montanus* (circles) as OTUs along PC1 and PC2 based on 12 quantitative characters (symbols as in Table 2) (after Kucharczyk & Kucharczyk, 2009)

Features	Female	Male
length of pronotal posteroangular seta interna	A	A
length of pronotal posteroangular seta externa	B	B
length of pronotal posteromarginal median seta	C	C
number of discal setae on sternum V	D	D
number of discal setae on sternum VII	E	E
number of discal setae on sternum VIII	-	F
max. number of distal setae on first vine of forewing	G	G
width of pore plate on sternum V	-	H
width of pore plate on sternum VII	-	I
length of antennal segment VII	-	J
length of antennal segment VIII	-	K
length of antennal segment III	M	M
max. length of median microtrichia on tergum VIII	O	-

Table 2. Features of *Thrips atratus* and *Thrips montanus* adults used in quantitative analyses (after Kucharczyk & Kucharczyk 2009)

2.4 Identification of *Thrips fuscipennis* Haliday 1836 and *T. sambuci* Heeger 1854 species

Similar problem exist in distinguishing *Thrips fuscipennis* Haliday 1836 and *T. sambuci* Heeger 1854. The former is a polyphagous species most often feeding in flowers while the latter is a monophagous insect feeding, breeding and developing on the lower side of *Sambucus* spp. leaves. In spite of both of the species being recognized as valid, their adults may be distinguished mainly by differences in color of antennal segments (Schliephake & Klimt, 1979; zur Strassen, 2003). Due to the fact that the color characters are very variable in specimens originated from different populations and stations, it is not possible to accurately identify species by using only their color features (Strassen zur, 1997; Mound & Minaei, 2010).

The aim of this task was to find new characters and test their usefulness in distinguishing these taxa. Fedor et al. (2008) proposed to use an artificial neural network method (ANN) for identifying species. This method was successfully applied according to 18 species of four genera: *Aeolothrips*, *Chirothrips*, *Dendrothrips* and *Limothrips*. Finally the authors selected 19 morphometric features which have been used in ANN analysis. In the current study seven of them were selected to distinguish *T. fuscipennis* and *T. sambuci*. Additionally five new quantitative features present in both sexes and three typical for males, and one qualitative feature were proposed for using in comparative study (Tab. 3).

Features	Female	Male
head width	H-w	H-w
head length (dorsal)	H-l-d	H-l-d
head length (ventral)	H-l-v	H-l-v
eye length	E-l	E-l
ovipositor length	Ov-l	-
antenal s. V length	A-V-l	A-V-l
antenal s. VI length	A-VI-l	A-VI-l
distance between anterior and posterior ocelli	D-oc	D-oc
distance between CS - metanotum	D-Cs-mt	D-Cs-mt
distance between D1 - metanotum	D-D1-mt	D-D1-mt
length of posteroangular seta interna	L-p-s-int	L-p-s-int
length of posteroangular seta externa	L-p-s-ext	L-p-s-ext
number of campaniform sensillae - mesonotum	N-cs-ms	N-cs-ms
distance between setae D1 and fore edge of metanotum	D-D1-e-mt	D-D1-e-mt
width of area porosae on sternum V	-	W-ap-sV
width of area porosae on sternum VI	-	W-ap-sVI
width of area porosae on sternum VII	-	W-ap-sVII

Table 3. Features of *Thrips fuscipennis* and *Thrips sambuci* adults used in quantitative analyses

Similarly as in *T. atratus* and *T. montanus* characters of females and males were analyzed separately. On the graphs being the result of PCA method the specimens of both species are well separated and are located on opposite sides of Axis 2 (Figs 12, 13). The number of campaniform sensilla on the mesonotum – two (sporadically one) in *T. fuscipennis* and lack of them in *T. sambuci* is the characteristic which in the highest degree differentiate these species. Additionally the measured setae are shorter in specimens of the former species. Males of the latter one are characterized by narrower area porosae on abdominal sternites V and VI and very often lack of them on sternite VII. Moreover, this analysis shows low significance such characteristics in recognizing these species as: eye length, distance between ocelli and length of antennal segments. The characteristics mentioned above were used in ANN analysis, the results obtained with PCA method tend to reflect on their usefulness in further similar studies, particularly on *Thrips* species identification.

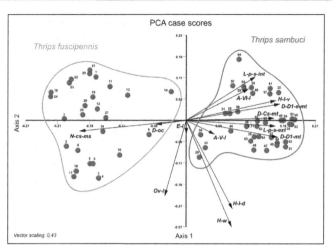

Fig. 12. PCA scatter diagram of female specimens of *Thrips fuscipennis* and *Thrips sambuci* as OTUs along PC1 and PC2 based on 14 quantitative characters (abbreviations as in Table 3)

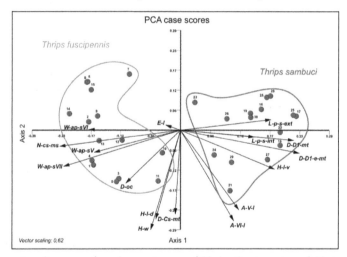

Fig. 13. PCA scatter diagram of male specimens of *Thrips fuscipennis* and *Thrips sambuci* as OTUs along PC1 and PC2 based on 16 quantitative characters (abbreviations as in Table 3)

3. Conclusion

The above mentioned examples show that PCA method is a valuable tool in identifying species. Its' application allows to select morphometric and qualitative characteristics which discriminate taxa and may be useful in construction of identification keys. This method also allows to verify the significance of some characteristics in taxonomic study and select these most relevant from large set of data.

The results obtained using the numerical taxonomy methods could objectively reflect the taxonomic position of studied taxa.

4. Acknowledgements

Financial support for this project was partially provided by the Slovak Science Foundation VEGA 1/0137/11.

5. References

Apuan D.A.; Torres M.A.J. & Demayo C. G. (2010). Describing variations and taxonomic status of earthworms collected from selected areas in Misamis Oriental, Philippines using principal component and parsimony analysis. *Egyptian Academic Journal of Biological Sciences*, B Zoology, Vol. 2, No. 1, pp. 27-36, ISSN 2090-0759

Bhatti J.S. (1988). The orders Terebrantia and Tubulifera of the superorder Thysanoptera (Insecta). A Critica Appraisal, *Zoology (Journal of Pure and Applied Zoology)*, Vol. 1: pp. 167-240, ISSN 0970-1516

Bhatti J.S. (2005). Fifteen new families in the Order Terebrantia (Insecta). *Thysanoptera*, Vol. 1, pp. 27-49.

Bhatti J.S. (2006). The classification of Terebrantia (Insecta) into families. *Oriental Insects*, Vol. 40, pp. 339-375, ISSN 0030-5316

Chiapella J. (2000). The Deschampsia cespitosa complex in central and northern Europe: morphological analysis. *Botanical Journal of the Linnean Society*, Vol.134. No. 4, (August 2006), pp. 495-512, ISSN 1095-8339

Crespi, B.J.; Morris, D.C. & Mound, L.A. (2004). *Evolution of Ecological and Behavioural Diversity: Australian Acacia Thrips as Model Organisms*. Australian Biological Resources Study & Australian National Insect Collection. ISBN: 9780975020616, Canberra, Australia

Fedor P.; Malenovský J.; Vaňhara J.; Sierka W. & Havel J. (2008). Thrips (Thysanoptera) identification using artificial neural networks. *Bulletin of Entomological Research*, Vol. 98, No.5, (April 2008), pp. 437-447, ISSN 0007-4853

Hotelling, H. (1933). Analysis of a complex of statistical variable into principal components. *Journal of Educational Psychology*, Vol. 24, No. 6, (September 1933), pp. 417-441, ISSN 0022-0663

Kovach Computing Services (2005). *Multi-Variate Statistical Package Plus, Version 3.1*. Kovach Computing Services, Pentraeth, Wales, U.K.: 137 pp.

Kucharczyk H. (2004). Larvae of the Genus Thrips – Morphological Features in Taxonomy. *Acta Phytopathologica et Entomologica Hungarica*, Vol. 39 No. 1-3, (May 2004), pp. 211-219, ISSN 0238-1249

Kucharczyk H. (2010). *Comparative morphology of the second larval instar of the Thrips genus species (Thysanoptera: Thripidae) occurring in Poland*. Mantis Publishing Company, ISBN 978-83-929997-7-5, Olsztyn, Poland

Kucharczyk H. & Kucharczyk M. (2009). *Thrips atratus* HALIDAY, 1836 and *Thrips montanus* PRIESNER, 1920 (Thysanoptera: Thripidae) – one or two species? Comparative morphological studies. *Acta Zoologica Academiae Scientiarum Hungaricae*, Vol. 55, No. 4, (November 2009), pp. 349-364, ISSN 1217-8837

Lilburn T.G. & Garrity G.M. (2004). Exploring prokaryotic taxonomy. *International Journal of Systematic and Evolutionary Microbiology*, Vol. 54, No. 1, (January 2004), pp. 7-13, ISSN 1466-5026

Lewis T. (Ed.). (1997). *Thrips as Crop Pests*. CAB International, ISBN 0 85199 178 5, Wallingford, UK

Mound L.,A.; Heming C., S. & Palmer J., M., (1980). Phylogenetic relationships between the families of recent Thysanoptera (Insecta). *Zoological Journal of the Linnean Society*, Vol. 69, No. 2, (June 1980), pp. 111-141, ISSN 1096-3642

Mound L.A. & Morris D.C. (2004). Thysanoptera Phylogeny – the Morphological Background. *Acta Phytopathologica et Entomologica Hungarica*, Vol. 39, No. 1-3, (May 2004), pp. 101-113, ISSN 0238-1249

Mound L.A. & Morris D.C. (2007). The insect order Thysanoptera: Classification versus Systematics. *Zootaxa*, Vol. 1668, (December 2007), pp. 395-411, ISSN 1175-5326

Mound L.A. (2010). Classifying thrips into families. http://anic.ento.csiro.au/thrips/ identifying_thrips/classification.html [accessed October 2011].

Mound, L. A., Minaei, K. (2007). Australian insects of the Haplothrips lineage (Thysanoptera – Phlaeothripinae). *Journal of Natural History*, Vol. 41, pp. 2919-2978, ISSN 0022-2933

Mound, L.A. & Minaei, K. (2010). Taxonomic problems in character state interpretation: variation in the wheat thrips *Haplothrips tritici* (Kurdjumov) (Thysanoptera, Phlaeothripidae) in Iran. *Deutsche Entomologische Zeitschrift*, Vol. 57, No. 2, (November 2010), pp. 233-241, ISSN 1435-1951

Pearson, K. (1901). On lines and planes of closest fit to systems of points in space. *Philosophical Magazine*, Vol. 2, pp. 559-572, ISSN 1478-6443 http://pbil.univ-lyon1.fr/R/pearson1901.pdf [accessed October 2011].

Pitkin, B. R. (1976). A revision of the Indian species of *Haplothrips* and related genera (Thysanoptera: Phlaeothripidae). *Bulletin of the British Museum (Natural History), Entomology series*, Vol. 34, pp. 223-280, ISSN 0524-6431

Sahuquillo E.; Fraga M.I. & Martinez Cortizas A. (1997). Comparative study of classical and numerical taxonomic methods for infraspecific taxa of *Triticum aestivum* L. traditionally cultivated in Galicia (NW Spain), pp. 919-926. Valdes B. & Pastor J. (Eds.), *Proceedings of the VIII optima meeting*. Sevilla 25 Sep.-1 Oct. 1995.

Schliephake G. (2001). Verzeichnis der Thysanoptera (Fransenflügler) – Physopoda (Blasenfüße). [In] *Entomofauna Germanica* Klausnitzer B. (Ed.).Vol. 5, pp. 91-106, ISSN 0232-5535

Schliephake G. & Klimt K. (1979). *Thysanoptera, Fransenflügler*. Die Tierwelt Deutschlands, Vol. 66. Veb Gustav Fischer Verlag Jena, Germany

Strassen zur R. (1997). How to classify the species of the genus *Thrips* (Thysanoptera)? *Folia Entomologica Hungarica*, Vol. 58, pp. 227–235, ISSN 0373-9465

Strassen zur R. (2003). *Die terebranten Thysanopteren Europas und des Mittelmeer-Gebietes*. [In:] Die Tierwelt Deutschlands Vol. 74 (Goecke & Evers eds). ISBN 3-931374-58-0, Keltern, Germany

Tommasini M. G. & Maini S. (1995). Frankliniella occidentalis and other thrips harmful to vegetable and ornamental crops in Europe, In: *Biological control of thrips pests* Loomans A., J.; M., van Lenteren J.,C.; Tomasini M., G.; Maini S. & Riudavets J. (Eds.), pp. 1-42, Wageningen Agricultural University Papers, 95.1, ISBN 90-6754-395-0, Wageningen, The Netherlands

Wolff R. L.; Comps B.; Marpeau A. M. & Deluc L.G. (1997). Taxonomy of *Pinus* species based on the seed oil fatty acid compositions. *Trees – Structure and Function*, Vol. 12, No. 2, (December 1977), pp. 113-118, ISSN 0931-1890

Véla E.; Tirard A.; Rinucci M.; Suesh C.M. & Provost E. (2007). Floral Chemical Signatures in the Genus *Ophrys* L. (Orchidaceae): A Preliminary Test of a New Tool for Taxonomy and Evolution. *Plant Molecular Biology Reporter*, Vol. 25, No. 3-4, (December 2007), pp. 83-97, ISSN 0735-9640

Zawirska I. & Wałkowski W. (2000). Fauna and importance of Thrips (Thysanoptera) for rye and winter wheat in Poland. *Journal of Plant Protection Research*, Vol. 40, No. 1, pp. 35–55, ISSN 1427-4345

7

Application of the Principal Component Analysis to Disclose Factors Influencing on the Composition of Fungal Consortia Deteriorating Remained Fruit Stalks on Sour Cherry Trees

Donát Magyar[1] and Gyula Oros[2]
[1]National Institute of Environmental Health, Budapest,
[2]Plant Protection Institute of the Hungariain Academy of Sciences, Budapest,
Hungary

1. Introduction

The sour cherry production is concentrated in eastern region of Hungary. During last two decades the commercialized yield varied between 42000-90000 metric tonnes. One of the main constraints is the fruit decay (anthracnose) caused by *Colletotrichum gloeosporioides* (Penz.) Penz. & Sacc. in Penz. (teleomorph: *Glomerella cingulata* [Stoneman] Spauld. & H. Schrenk). This disease has been known for centuries, and in certain orchards yield losses greater than 90% periodically occurred under epidemic conditions. In last decade, however, the grave infection has evolved in each season that resulted in disastrous yield losses and led to decrease of harvested area at about 25 %. Series of control measures were tried out, many of them with success with some degree (Børve & Stensvand, 2006; Børve et al., 2010), among them a newly developed biocontrol preparation was applied (Oros & Naár, 2008; Oros et al., 2011). Detailed analysis of weak efficiency of disease control revealed that beside *C. gloeosporioides* four other anamorphs of *Glomerella* were present (Table 1), among them *C. acutatum* (J.H. Simmonds, 1968), - a new pathogen for the region, - became recently dominant. Its strains, tolerant to recently applied fungicides to control the anthracnose, could be isolated of sour cherry orchards that might be the cause of ineffectiveness of control measures in 2006-2010 (Oros et al., 2010). At the increasing costs of cultivation only high yields may secure a profitable production of sour cherry. From this point of view, the anthracnose caused by *Glomerella* anamorph with special regards to *C. acutatum* is an important element of uncertainty that means elaboration of efficient control measures is the critical challenge to unconquerable sour cherry production. Efficient control measures reduce the yield loss in the orchard by suppressing the pathogen (i.e. by the reduction of the number of viable spores and infected fruits). On our major surprise, the pathogen can not be isolated of overwintering fruit mummies in industrial plantations of East Hungary contrarily to other sour cherry producing regions. A survey was carried out to disclose the cause of this phenomenon collecting samples of remained fruit stalks of varieties in sour cherry gene bank of East Hungary. The relationships between host (sour cherry variety) and fungi associated to stalks have been analyzed by multivariate methods: Non-linear Mapping and Principal Component Analysis (PCA) combined with Potency Mapping (PM) and Spectral Mapping (SPM) techniques.

	Location	Cglo	Cacu	Ctru	Ctri	n.d.
Újfehértó	47°46′46.7″N, 21°39′10.2″E	+	+	+	-	+
Újfehértó	47°40′22.2″N, 21°41′24.7″E	+	+	-	-	-
Pilis	47°15′45.9″N, 19°34′13.1″E	+	+	+	+	+
Szentendre	47°44′94.3″N, 19°03′10.7″E	+	-	-	-	-
Budapest	47°31′34.3″N, 19°01′44.7″E	+	+	-	-	-
Budapest[b]	47°30′50.8″N, 19°00′38.5″E	-	-	-	+	-
Érd	47°24′07.3″N, 18°55′26.1″E	-	+	-	-	+
Agárd	47°11′03.8″N, 18°35′15.1″E	+	+	-	-	+
Szombathely[c]	47°13′51.3″N, 16°33′33.8″E	-	+	-	-	+
Jakabszállás	47°45′16.3″N, 14°37′24.4″E	+	+	-	-	-
Kecskemét	46°53′45.9″N, 19°42′24.8″E	+	+	-	-	-
Jakabszállás	46°29′28.7″N, 19°16′22.7″E	+	+	-	-	-

Table 1. Presence of *Colletotrichum* (Corda) species related to sour cherry in various regions of Hungary[a].

[a]=Isolates of each species gangrene the fruits of sour cherry (*Prunus cerasus* L. cv. Kántorjánosi), [b]=collected of *Hedera helix* L., [c]= herbarium specimen of *Fallopia sachalinensis* (F. Schmidt) Ronse Decr.; Cglo=*C. gloeosporioides* (Penz.) Penz. & Sacc. (*Glomerella cingulata* (Stoneman) Spauld. & H. Schrenk); Cac=*C. acutatum* J.H. Simmonds (*G. acutata* Guerber & J.C. Correll), Ctrun=*C. truncatum* (Schwein.) Andrus & Moore (*G. truncata* C.L. Armstr. & Banniza), Ctri=*C. trichellum* (Fr.) Duke; n.d.=unidentified.

Various multivariate mathematical–statistical methods have been widely applied to elucidate the relationship between the columns and rows of any data matrix. In general, the choice of a particular multivariate method for data analysis is dictated by the properties of the data and by the objectives. The PCA, introduced by Pearson (1901), is a useful and widely applied technique to reduce the number of multidimensional contrasts and to reproduce the original data in low dimensional space. In this perspective PCA is a versatile and an easily applicable multivariate method, which forms clusters of variables and observations taking into consideration simultaneously the overall effect of factors under investigation without being one of dependent variable (Mardia et al., 1979). In last four decades over 43000 scientific papers were published that applied PCA for data analysis among them more than 800 intensively reviewed them (Fig 1). There are large literatures in diverse areas such as psychometrics, chemometrics, computer science and ecology applying PCA. Biological and environmental sciences are among the top users. Such statistic is widely used in these works to determine the source of a variety of soil, plant and atmospheric contaminants, e.g. heavy metals (Titseesang et al., 2008; Schwarz et al., 1999; Tokalioglu et al ., 2010), trace metals (Rungratanaubon et al., 2008; Shaheen et al., 2005; Shah et al., 2006; Vasconcellos et al., 2007), PM 2.5 and PM10 particles (Jacquemin et al., 2009; Quraishi et al., 2009) and volatile organic compounds (You et al., 2008).

Especially in agricultural sciences, where complex datasets with biotical and abiotic variables are often used, the application of multivariate statistics is fundamental. In these studies, PCA is mostly applied to reduce the number of input variables (Uno et al., 2005). Also, these calculations are used to develop precision agriculture applications using crop monitoring (Yang et al., 2004), to uncover gradients of canopy structure (Frazer et al., 2005), to determine the time of harvest (Garcia-Mozo et al., 2007) or to evaluate sampling methods (Magyar et al., 2011). Although, in particular, atmospheric science is a rich source of both applications and methodological developments that influenced applications of PCA. In the field of aerobiology relatively few papers have been published. PCA was applied to analyze the variability of fruit production based on aerobiological pollen data (Ribeiro et al., 2007, 2008), as well as to connect meteorological parameters with the periods of high pollen concentrations (Parrado et al., 2009). This method proved to be reliable to identify the sources and dispersal patterns of airborne bacteria (Lighthart et al., 2009) and spores of plant pathogenic fungi (Frank et al., 2008; Magyar, 1998) that resulted the stressing of its possible use for both signalization of occurrence and identification of sources of plant pathogens. The latter is important in relation to successful control measures in the agricultural areas as the knowledge on the type of dispersal of the pathogen (and biocontrol agent if applied) and the composition of fungal consortia on the target area is crucial for an effective pest control (Magyar 2007).

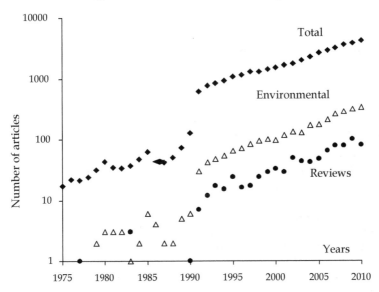

Fig. 1. Changes in abundance of scientific publications that have applied Princpal Component Analysis. Source: Thompson Data Base

The PCA is an exploratory tool to investigate patterns in the data. Usually, the first Principal Component (PC) is accepted to comprise the summarized quantitative aspects of multifactorial effects as the arithmetic mean values of observations correlate with first PC, while the remaining PCs are considered to relate to spectrum of effects. However, the quantitative moments of PC variables are distributed within hidden factors, thus PCA cannot be employed when the separation of the strength and selectivity of the effect is required. The spectral mapping technique (SPM) was developed to overcome this difficulty (Lewi, 2005). Using the normalized original data this method divides the information into two matrices. The first one is a vector containing so-called potency values (PV) proportional to the overall effect of factors or the response of organism to them (quantitative measure of the effect). The second matrix (selectivity or spectral map) contains the information related to the variation in manifestation (spectrum) of the effect (qualitative characteristics of the effect). In our case PV relates to the overall abundance of each fungal species noticed on stalks, i.e., it is a quantitative measure of the diversity of fungal consortium, while the spectral map (SM) comprises qualitative characteristics of variety dependent distribution of species in the orchard. The matrices of PC loadings and variables as well as the SM are generally multidimensional. As the human brain is not suitable for the evaluation of the distribution of data in multidimensional space a nonlinear mapping technique (NLMAP) was developed for the reduction of the dimensionality of such matrices (Sammon, 1969).

Our objectives of this study were to evaluate diversity of mycobiota associated to stalks of sour cherry remained on trees, and in this paper we show the use of PCA in conjunction with SPM and NL-Mapping to reveal patterns in distribution of detected fungal species.

2. Experimental

2.1 Location and method of sampling

2.1.1 Sour cherry fruits were collected in ten different locations (Table 1) of Hungary in period of fruit ripening during 2006-2010 and searched for *Colletotrichum* acervuli immediately after taking them into the laboratory. The *Colletotrichum* species are anamorphs of some *Glomerella* (Ascomycota) fungi and their presence was proved microscopically by usual manner. Of the same locations twig s with 15-25 buds and mummies with stalks were collected in February of subsequent years.

2.1.2 The sour cherry gene bank is located in Újfehértó (East Hungary, 47°46′46″N, 21°39′10″E), and it is surrounded by big sized industrial plantations that comprise mainly three varieties (*Prunus cerasus* L. cv. Érdi bőtermő, Kántorjánosi and Újfehértói fürtös). The up to date management practices have usually been carried out in the area. On the nearby two plots (4 ha size of each) *Trichoderma*-based biopreparations (Oros and Naár, 2008) were applied during past three vegetation periods (Oros et al., 2010, 2011). Fruit stalks remained on trees were collected in March 2011 (10 samples of each tree and 10-30 stalks per sample) of 18 varieties of gene bank as well as of trees of the two industrial plots and kept in freezer up to analysis.

2.2 Survey of fungal population on stalks

2.2.1 The twigs collected in diverse areas in first week of February (2.1.1.) were plug in pots filled with wetted perlite and tucked into plastic bag for 48 hours at ambient temperature to promote acervuli formation on surface of tissues infected with *Colletotrichum*. The acervulus is a characteristic reproductive organ of *Glomerella* anamorphs that easy to recognize of dark brown, thorn like, prominent setae examining samples under dissecting microscope.

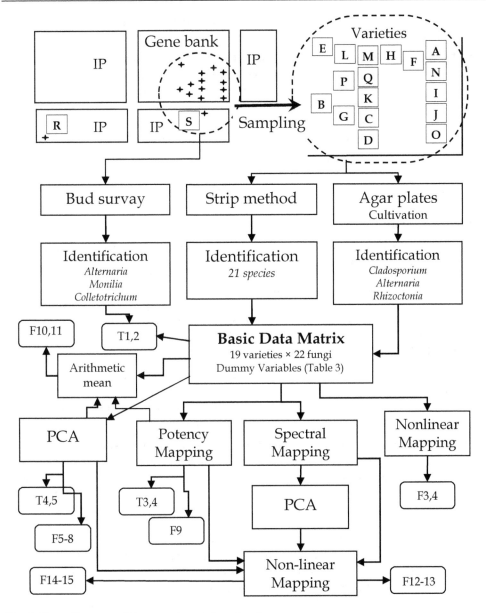

Fig. 2. Flow diagram of the experimental protocol.

IP= industrial plantation. The capital letter codes of varieties are given in Table 2.
The labels in ellipses (T1,2..n) and (F1,2..n) mark number of tables and figures in body text where the
results of computation was used for demonstration.

2.2.2 Stalks collected of trees in winter were placed on the surface of Czapek Agar plates
amended with antibiotics (ampicillin, kanamycin and streptomycin) to supress bacteria. The

growth of fungi on the surface of stalks as well as on the agar plates had been survayed daily during three weeks.

2.2.3 To study the fungal spectra of the fruit stalks, a microscopic method was applied which was developed to enable *in situ* investigations by the use of pressure sensitive acrylic strips (MACbond B 1200, MACtac Europe S.A., Brussels). The strip consists of a thin (20 µm) polypropylene film coated on both sides with a rubber based adhesive, which offers very high adhesion property on the stalks. The tape is viewed directly under a microscope to identify the fungi present on the sampled surface under ×800 magnification.

All mycological work was done applying usual manners widely accepted in microbiology. Spores have been grouped into hydrophobic dry spores (xerospores) and readily wettable slime spores (gloiospores) (Mason, 1937, Magyar, 2007).

2.3 Data treatment

The basic data matrix consisted of the dummy variables that comprise presence (1) or absence (0) of fungal species (n=22) on the stalks of the sour cherry variety (n=19) concerned, which strongly limits the use of data transformation procedures. The yeast species, which occurred on each stalk examined, had been excluded of further data analyses, because PCA and SPM can be employed for the analysis of data matrices which contain variables showing variance. First the Nonlinear Mapping (NLMAP) was applied for clustering both sour cherry varieties and fungi detected on stalks following Sammon (1969).

PCA was carried out on the correlation matrix and only the components having an eigenvalue greater than one were included in the evaluation of data. To facilitate the evaluation of the multidimensional maps of principal component (PC) loadings, their dimensionality was reduced to two by the NLMAP technique.

Potency and Spectral Mapping techniques (Lewi, 2005) were applied on the matrix used for PCA, and applied in order to calculate the abundance and the variety dependent distribution of fungal species as well as the degree of diversity and the heterogeneity spectrum of settlement on proper cherry variety. Potency and Spectral Mapping were carried out twice: (a) The abundance of 19 fungal species represented the variables and the 19 sour cherry variables the observations (spectral map calculated comprises the heterogeneity spectrum of fungal settlement on varieties). (b) The cherry varieties represented the variables and the abundance of 19 fungal species the observations (spectral map calculated comprises the variety dependent distribution of fungi). The potency values calculated relate to either the abundance of fungal species in sour cherry plantation or the degree of diversity of fungal consortium found on stalks of proper variety. The dimensionality of the maps was reduced to two by the NLMAP technique or hidden variables were extracted by PCA as the evaluation of the multidimensional selectivity (spectral) maps of varieties and associated fungi is difficult.

Statistical functions of Microsoft Office Excel 2003 (Microsoft, Redmondton, USA) and Statistica5 program (StatSoft, Tusla, USA) were used for analysis of data. The graphical presentation of result of data analysis was edited uniformly in MSOffice Power Point 2003.

Sour cherry varieties			Fungal species detected on stalk remained on trees		
No.	Code[a]	Gene bank	No.	Code	Taxonomic position[c] and scientific name
1	A	Aranka			Ascomycota
2	B	Bigarreau			Pezizomycotina
3	C	Bosnyák	1	tol	*Trullula olivascens* (Sacc.) Sacc.
4	D	Csengődi			Dothideomycetes, Botryosphaeriales
5	E	Debreceni b.	2	dte	*Diplodia tecta var. cerasii* Berk. & Broome
6	F	Érdi bőtermő			Dothideomycetes, Capnodiales
7	G	Géczi	3	cla	*Cladosporium sp.*
8	H	Germersdorfi			Dothideomycetes, Pleosporales
9	I	Halka	4	agl	*Aglaospora profusa* (Fr.) De Not.[g]
10	J	Horka	5	alt	*Alternaria sp.*
11	K	Katalin	6	ppo	*Phoma pomorum* Thüm.
12	L	Maliga emléke	7	epu	*Epicoccum purpurascens* Ehrenb.
13	M	Margit	8	lep	*Leptosphaeria sp.*
14	N	Pándy 56	9	phe	*Pleospora herbarum* P. Karst.
15	O	Sylvana	10	ple	*Pleospora sp.*[g]
16	P	Szomolyai	11	sfo	*Splanchnonema foedans* (Fr.) Kuntze[g]
17	Q	Vanda	12	ste	*Stemphylium sp.*
					Sordariomycetes[c]
	Industrial plantation[b]		13	sor	unidentified sp.[d]
					Sordariomycetes, Glomerellales[e]
18	R	KántorjánosiU	14	cgl	*Colletotrichum gloeosporioides* (Penz.) Penz. & Sacc.
19	S	KántorjánosiH			Sordariomycetes, Xylariales
			15	dpl	*Dendrophoma pleurospora* (Sacc.) Sacc.
					Sordariomycetes, Diaporthales
			16	cyt	*Cytospora sp.*
					Pezizomycotina, Helotiales
			17	mon	*Monilia sp.*
			18	m-s	Steril mycelium[f]
			19	y-w	yeast 1 – white
			20	y-r	yeast 2 – red
			21	y-b	yeast 3 – black
					Basidiomycota
					Agaricomycetes, Cantharellales
			22	rhi	*Rhizoctonia sp.*

Table 2. Stalk samples collected in gene bank of Újfehértó and the list of detectable fungi associated to.

[a]=The codes of sour cherry varieties (capital letters) and fungal species colonized them during the winter are used in following tables and figures. [b]=These plantations are adjacent to the gene bank. [c]=Species were assigned according to Index Fungorum. [d]=Detected only on a single stalk, [e]=Immature perithecia, [f]=Mycelium of an ascomycetaceous fungal species, [g]= free spores.

3. Results

3.1 The presence of *Colletotrichum* species was in each location stated on ripening fruits during vegetation periods of 2006-2010, and among them *C. acutatum* proved to be dominant being followed by *C. gloeosporioides* (Table 1). The occurrence of other *Glomerella* anamorphs was sporadic. In several samples acervuli of a *Colletotrichum* species unknown for Hungarian mycoflora were observed. Neither on leaves collected during vegetation period nor on decaying fruits were found *Glomerella* perithecia (sexual reproductive organ) in 2006-2011. The yield losses caused by anthracnose were significant in each location. The role of other pathogenic fungi was insignificant, among them *Monilia* sp. was most frequent, although in few cases *P. pomorum*, *D. tecta* and *Gibberella avenacea* R.J. Cook associated also to fungal consortium. The sporadic appearance of other than *Colletotrichum* fungal species was seemingly connected to intensive pest management in sour cherry plantations.

3.2 On twigs collected in February of 2007-2011 the rate of infection varied between 2 and 90 percents (method 2.2.1). *C. acutatum*, *C. gloeosporioides* and *C. trichellum* could be detected (Table 1) in listed order, but the samples of Újfehértó where *C. gloeosporioides* was only found. In the gene bank (L-1) the manifested infection rate on the twig s was low, moreover, few infected buds could be found only on two varieties (*P. cerasus* L. cv. Érdi bőtermő and Újfehértói fürtös). Also, *C. gloeosporioides* was the single overwintering *Colletotrichum* in Central Hungary (L-4), although in this orchard acervuli of other *Glomerella* anamorphs were not discovered on decayed fruits in vegetation period. *C. trichellum* was detected in one orchard only (L-3) on fruits, and in the same place it was found on basal parts of stalks. *C. truncatum* was found only on fruits in two locations (L1 and L3) suggesting that this species does not overwinter on sour cherry trees. Other fungal species were detected such as *Alternaria spp.*, *Cytospora* sp. and *Phoma pomorum*.

3.3 Counteracting to special analysis cosmopolite fungi *Alternaria* and *Cladosporium* overgrew rapidly all others on stalks placed on the Czapek Agar plates (method 2.2.2) except *Rhizoctonia solani* J.G. Kühn (teleomorph: *Thanatephorus cucumeris* [A.B. Frank] Donk) that was identified of two varieties (Pándy 56 of gene bank and Kántorjánosi of industrial plantation). The *Trichoderma* strains of biopreparate were not detectable. The acervulus of *C. gloeosporioides* could be detected on a single stalk (variety Aranka of gene bank).

3.4 Following protocol 2.2.3. we could detect presence of 8 known phytopathogenic and 13 saprobiont fungal species in samples collected in gene bank of Újfehértó (Table 2). Five species occurred only in a single variety (Table 3). Thus, intensively growing sterile mycelium of an ascomycete was observed on stalks of Debreceni bőtermő (N, letter codes of varieties are given in Table 2); however, this did not counteract the detection of other fungal species. Also, immature perithecia could be seen on stalks of Géczi (E). Spores of three phytopathogenic species causing branch dieback, *Leptosphaeria*, *Splanchonema* and *Cytospora* were detected on stalks of varieties B, P and R, respectively. The most frequent *Alternaria* and *Stemphylium* being facultatively parasitic can take part in fruit decay. The occurrence of *Monilia*, one of the most devastating pathogens of sour cherry, was less pronounced as expected. The other species are known as saprobionts and we have no information about their possible phytopathogenic properties, nevertheless, these can also have importance in postharvesting fruit rot. Presence of none of *Glomerella* anamorph but *C. gloeosporioides* on some stalks of a single variety (A) could be detected.

Application of the Principal Component Analysis to Disclose Factors Influencing on the Composition of Fungal
Consortia Deteriorating Remained Fruit Stalks on Sour Cherry Trees

113

Fungi[a]							Sour cherry varieties[a]													
Code	A	B	C	D	E	F	G	H	I	J	K	L	M	N	O	P	Q	R	S	PV[b]
tol	0	0	1	0	0	1	0	1	1	0	0	1	1	1	1	1	0	0	1	52.6
dte	0	0	0	0	0	0	0	0	0	0	0	0	0	1	1	0	0	0	0	10.5
cla	1	0	0	0	1	0	0	0	0	0	0	0	0	0	1	0	1	0	1	26.3
agl	0	0	1	0	0	0	0	1	0	0	0	0	0	0	0	0	0	1	1	21.1
alt	1	1	1	0	1	1	1	1	1	1	1	1	1	1	1	1	1	1	1	94.7
ppo	0	0	1	0	0	1	0	0	0	0	0	0	0	1	1	0	0	0	0	21.1
epu	0	0	0	0	0	0	0	0	0	0	0	0	0	0	1	1	1	0	0	15.8
lep	0	1	0	0	0	0	0	0	0	0	0	0	0	0	0	0	0	0	0	5.3
phe	1	0	0	1	1	0	0	0	0	0	1	0	0	0	1	0	0	1	0	31.6
ple	0	0	0	0	1	0	0	0	0	0	0	0	0	0	0	1	0	0	0	10.5
sfo	0	0	0	0	0	0	0	0	0	0	0	0	0	0	0	1	0	0	0	5.3
ste	1	0	0	1	1	0	0	0	1	1	1	0	0	1	1	1	1	0	0	52.6
sor	0	0	0	0	0	0	1	0	0	0	0	0	0	0	0	0	0	0	0	5.3
cgl	1	0	0	0	0	0	0	0	0	0	0	0	0	0	0	0	0	0	0	5.3
dpl	0	0	1	0	0	0	0	0	0	0	1	0	0	0	1	0	1	0	0	21.1
cyt	0	0	0	0	0	0	0	0	0	0	0	0	0	0	0	0	0	1	0	5.3
mon	0	1	0	0	0	0	1	0	0	0	0	0	0	1	0	0	0	0	1	21.1
ms	0	0	0	0	1	0	0	0	0	0	0	0	0	0	0	0	0	0	0	5.3
y1	1	1	1	1	1	1	1	1	1	1	1	1	1	1	1	1	1	1	1	100
y2	1	1	1	1	1	1	1	1	1	1	1	1	1	1	1	1	1	1	1	100
y3	1	1	1	1	1	1	1	1	1	1	1	1	1	1	1	1	1	1	1	100
rhi	0	0	0	0	0	0	0	0	0	0	0	0	0	1	0	0	0	0	1	10.5

Table 3. Mycological analysis of stalks collected in gene bank of Ujfehértó.

[a]=codes of fungi and varieties are shown in Table 2. [b]=Potency Values as percent of the maximum abundance. 0=not found, 1=presented on more than 1 specimen.

3.5 Data analysis

3.5.1 Non-linear mapping

Both sour cherry varieties and associated fungal species were mapped following method of Sammon (1969). The fungi might be grouped by their relation to host plant, one of clusters comprise pathogens (dotted line) while the other the saprobionts mainly (Fig 3).

In the case of varieties no clusters could be recognized (Fig 4). The four varieties (E, F, S and R) are near relatives, and their position on NLMap is not linked. These were selected for free pollination and improving the mechanic harvest. Unfortunately, data on the genetic background of other varieties is absent; the varieties B, O and Q were introduced of other geographic regions. The varieties F, R, S and E are extremely susceptible to anthracnose caused by C. acutatum in decreasing order; moreover their susceptibility to Monilia is outstanding too. Nevertheless, due to the high commercial value of their fruits and favorable biological properties these are the most run varieties in Hungary.

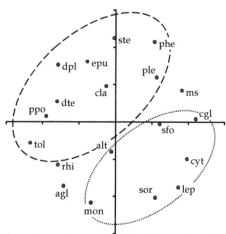

Fig. 3. Non-linear Map of fungal species as variables. The codes of fungi are given in Table 3. Ellipses with dotted and broken lines mark groups of pathogenic and saprotrophic fungi, resp.

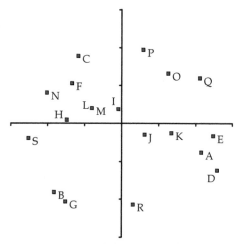

Fig. 4. Non-linear Map of sour cherry varieties as variables. The codes of varieties as given in Table 3.

3.5.2 Principal Component Analysis

3.5.3 Influence of sour cherry varieties on distribution of associated fungi

The PC loadings (sour cherry varieties as variables) computed by PCA are compiled in Table 4. Five hypothetical (background) variables explained the majority of total variance with 22.57 % loss of information. This fact suggests that the 19 original variables can be reduced to 5 theoretical variables. Most of varieties have high loadings in the first two PCs, and those being in gene bank are positively related, that means these two factors act similarly on distribution of associated fungi when deteriorating stalks of varieties concerned. The two sampling points of industrial plantations (R and S) have not common weight with the former group.

On bivariate plot of varieties as PC variables no connection can be observed similarly to NL-map (Fig 4), though these two plots are different. The distribution fungal species plotted as PC scores does not show connection between them, although the xerospore producing anemophilous species seem to deviate of the main group comprising gloeospore producing ones.

No.	Variables Sour cherry varieties	Code	PC1	PC2	PC3	PC4	PC5	PV[a]
1	Aranka	A	0.44	**0.65**	-0.22	0.14	0.01	36.4
2	Bigarreau	B	0.34	-0.23	**-0.69**	-0.25	0.15	27.3
3	Bosnyák	C	**0.60**	-0.40	0.32	0.36	0.31	36.4
4	Csengődi	D	0.20	**0.80**	-0.02	0.12	0.02	22.7
5	Debreceni bőtermő	E	0.35	**0.64**	-0.18	0.07	-0.29	40.9
6	Érdi bőtermő	F	**0.74**	-0.40	0.26	-0.01	0.12	27.3
7	Géczi	G	0.34	-0.23	**-0.69**	-0.25	0.15	27.3
8	Germersdorfi	H	**0.73**	-0.44	0.01	0.39	-0.18	27.3
9	Halka	I	**0.90**	0.09	0.09	-0.19	-0.18	27.3
10	Horka	J	**0.75**	0.43	-0.21	-0.21	-0.05	22.7
11	Katalin	K	**0.56**	**0.64**	-0.07	0.18	0.26	31.8
12	Maliga emléke	L	**0.88**	-0.35	0.06	0.01	-0.19	22.7
13	Margit	M	**0.88**	-0.35	0.06	0.01	-0.19	22.7
14	Pándy 56	N	**0.56**	-0.20	0.04	-0.44	0.35	45.5
15	Sylvana	O	**0.56**	0.35	0.49	-0.02	0.42	54.5
16	Szomolyai	P	**0.54**	0.12	0.28	-0.35	**-0.58**	40.9
17	Vanda	Q	0.47	0.49	0.12	-0.16	0.21	36.4
18	Kántorjánosi-U	R	0.33	0.07	-0.33	**0.78**	-0.06	31.8
19	Kántorjánosi-H	S	0.49	-0.44	-0.28	0.10	0.08	40.9
Eigenvalues (Explained variation)			6.74	3.54	1.79	1.51	1.13	
Proportion total			35.5	18.6	9.4	8.0	6.06	

Table 4. Principal Component Loadings

[a]=Potency Values as percent of the maximum abundance.

3.5.4 Similarities in distribution of fungi on stalks of sour cherry

Nine hidden (background) variables explained the 87.82 % of total variance, however, altering of the variations in susceptibility of sour cherry none of eigenvalues exceeded 20 %. Six variables (fungal species) dominated two PCs, while only three PCs were dominated by more then one variable on bivariate blots by two major PCs neither PC loadings (Fig 7, fungi) nor PC scores (Fig 8, varieties) are designated to explainable clusters, because neither taxonomic position nor mode of distribution of fungi could be linked to groups as well as relations of sour cherry varieties with similar genetic background (E, F, R and S) or their spatial position in gene bank could be linked on plot (Fig 8).

On our view the above facts indicate the multifactorial and highly specific character of interaction between sour cherry and associated mycobiota. This aspect seems to be support by formation of two big clusters on Figure 5, since genetic background of the mode of life is very complex and the strategy of infection of these pathogens is also different.

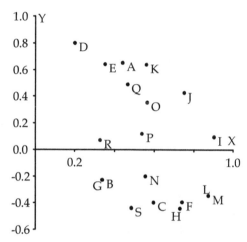

Fig. 5. Plotting sour cherry varieties as PC loadings by two major PCs. The capital letters mark varieties as given in Table 3. The distribution of variables is determined by 54 % of total variation.

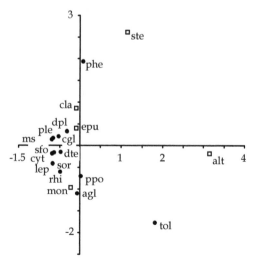

Fig. 6. Plotting fungal species as PC scores by two major PCs. The codes of fungi are given in Table 3. Full circles and opened squares mark species producing xero- or gloeospores, resp.

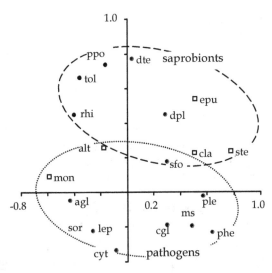

Fig. 7. Plotting fungal species as PC loadings by two major PCs components. The codes of
fungi are given in Table 3. Full circles and opened squares mark species producing xero- or
gloeospores, resp.

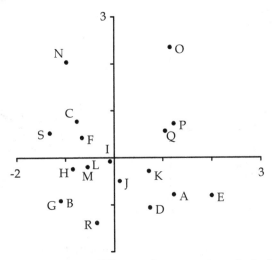

Fig. 8. Plotting sour cherry varieties as PC scores by two major principal components. The
capital letters mark varieties as given in Table 3. The distribution of variables is determined
by 33 % of total variation.

3.5.5 Potency mapping

The potency values (PV-SC) of individual sour cherry varieties and associated fungi
calculated by potency mapping technique are included into Tables 4 and 3; all PVs are
transformed into percents of the possible maximum for correlation analyses.

The potential diversity of mycobiota varied between 10-50 percents in gene bank (Fig 9A) that means no one of varieties was colonized with full spectrum of fungi detected in the area. The genetic background has seemingly minor importance in this respect because the PVs of related varieties are different. However, the highest PVs were measured one trees located near the edge of the orchard suggesting the importance of spatial position of trees influences in major extent the diversity of associated mycobiota than the genetic background. The potential incidence of fungal species (PV-FI) varied in higher extent (5-100 %) than PV-SC (Fig 9 B); being the yeasts the most dispersed fungi. The saprobionts had higher values than phytopathogens in general, although the facultatively parasitic *Alternaria* was dominant in majority of samples. The xerospore producing fungi were more frequent than the gloeospore-forming ones that can be explained by the season of sampling. Gloeospores, tend to be dispersed by water, e.g. rain splash (rain drops hit spore-filled water film on branches and leaves of trees, aerosolizing spores.) Occasionally, gloeospores are also dispersed by insects. In winter time both treetop and most of insect were absent.

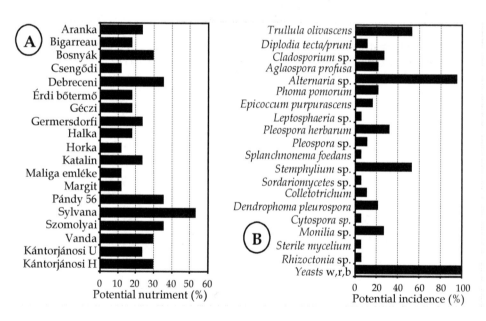

Fig. 9. Potency maps of sour cherry varieties and associated fungal species.

A – diversity of mycobiota associated to stalks takes maximum value in the case when all species detected in gene bank are represented in the sample
B – maximum incidence takes places when the species concerned is presented in all samples.

3.5.6 Regression analysis

The arithmetic means of columns and rows (host plant×fungi), potency values of varieties and fungi, the principal components scores of Tables 4 and 5 were correlated by linear regression analysis (Fig 10 and 11).

Variables[a]		Principal Components									Type[b]
No.	Code	PC1	PC2	PC3	PC4	PC5	PC6	PC7	PC8	PC9	
1	tol	-0.36	**0.66**	-0.29	0.31	0.17	-0.22	0.26	0.13	0.00	G
2	dte	0.03	**0.77**	0.30	-0.26	0.00	-0.15	-0.27	-0.12	0.00	G
3	cla	**0.51**	0.22	0.32	-0.32	0.38	0.36	0.24	-0.02	0.00	X
4	agl	-0.43	-0.05	0.19	0.46	**0.57**	0.14	-0.03	-0.14	0.00	G
5	alt	-0.18	0.26	-0.18	0.01	0.31	**0.64**	0.23	0.13	0.00	X
6	ppo	-0.16	**0.74**	0.26	0.10	-0.02	-0.10	-0.15	0.22	0.00	G
7	epu	**0.51**	**0.54**	-0.26	0.13	-0.18	0.37	-0.14	-0.30	0.00	X
8	lep	-0.26	-0.22	-0.11	-0.35	-0.22	0.29	-0.20	-0.02	-0.73	G
9	phe	**0.64**	-0.23	**0.55**	0.02	0.16	-0.12	-0.24	-0.13	0.00	G
10	ple	**0.57**	-0.02	**-0.66**	-0.11	0.41	-0.03	-0.16	0.07	0.00	G
11	sfo	0.30	0.18	**-0.79**	0.14	0.02	-0.02	-0.02	-0.43	0.00	G
12	ste	**0.78**	0.24	0.06	-0.24	-0.18	-0.25	0.01	-0.13	0.00	X
13	sor	-0.26	-0.22	-0.11	-0.35	-0.22	0.29	-0.20	-0.02	**0.73**	G
14	cgl	0.30	-0.19	0.33	-0.20	0.03	0.07	**0.69**	-0.30	0.00	G
15	dpl	0.29	0.45	0.36	0.29	-0.26	0.46	-0.18	0.17	0.00	G
16	cyt	-0.09	-0.34	0.25	0.42	0.38	0.09	-0.40	-0.43	0.00	G
17	mon	**-0.59**	0.09	-0.04	**-0.72**	0.08	0.12	-0.19	-0.21	0.00	X
18	ms	0.48	-0.20	-0.12	-0.30	**0.55**	-0.02	-0.20	**0.52**	0.00	u
19	rhi	-0.40	0.45	0.11	-0.44	0.43	-0.26	0.03	-0.25	0.00	u
Eigenvalue		3.38	2.83	2.18	1.95	1.66	1.32	1.19	1.12	1.06	
Prop. total		17.8	14.9	11.5	10.3	8.73	6.97	6.28	5.89	5.56	

Table 5. Principal component loadings

[a] = Codes are deciphered in Table 2. Yeast species were presented in each sample thus these three
variables did not show variance and were omitted of PCA.
[b] = Xerospores (X) mostly transported by wind, gloeospores (G) may be transported either by water of
animals.

The arithmetic mean on diversity of mycobiota of sour cherry varieties in our case correlated
only with potency values ($r^2=1$) but no correlation was revealed ($r^2=0.12$) with PC scores of
first PC and poor correlation ($r^2=0.49$) with PC scores of PC2 that well demonstrated on Fig
10. Contrarily, the arithmetic mean of dispersal significantly correlated both with potential
incidence of fungal species ($r^2=1$) and PC scores of first PC ($r^2=0.968$) and the relation with
scores of second PC is evidently non-linear (Fig11). This entire well demonstrates that the
first PC does not comprise always the majority of quantitative measures of the observations,
and it is necessary to check this relationship before using PC1 in further calculations.

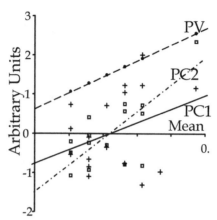

Fig. 10. Relationships between arithmetic mean of diversity of mycobiota and PVs and PC scores of sour cherry varieties.

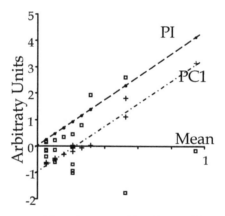

Fig. 11. Relationships between arithmetic mean of incidence of fungal species and PVs and PC scores.

3.5.7 Spectral component analysis

The two-dimensional non-linear selectivity map of sour cherry varieties is shown in Figure 12. This map shows similarities to Figure 4. However, in this case there is possible to distinguish a group of varieties where anemophilous fungi dominate the associated fungal consortium deteriorating the remained stalks on trees.

Carrying out the PCA with SPM when varieties serve as variables resulted five significant PC components indicating the elimination of part of total variance of basic data matrix by Potency Mapping. None of them was dominant and on the bivariate plots any patterns could be recognized in grouping of sour cherry varieties (these details are not shown). However, the Non-linear mapping of PC variables revealed three cluster (Fig 13), where varieties seemingly grouped according to composition of the fungal consortia. Moreover, this pattern is better pronounced than on Figure 12 supporting the advantage of combination of spectral mapping with PCA as well as the use of NL mapping for studying matrices comprising variables where the setting of observed phenomena influenced by multidunous complex of factors synchronously (environmental, ecological, genetic, etc).

The varietal distribution of fungal species is shown on two-dimensional non-linear selectivity map in Figure 14. The species frequently found on decaying fruits formed a separated cluster, while those parasiting on leaves and phloeme distribute without clear grouping.

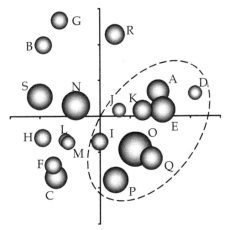

Fig. 12. Non-linear Map of sour cherry varieties as spectral variables. The capital letters mark varieties as given in Table 4. The size of balls is proportional to potential diversity of mycobiota recorded (Table 4). The ellipse marks the varieties colonized predominantly with xerospore producing fungi.

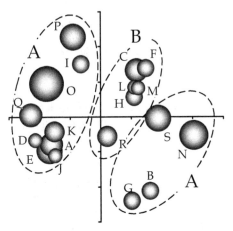

Fig. 13. Similarities in composition of mycobiota on stalk of sour cherry varieties. The capital letters mark varieties as given in Table 4. The size of balls is proportional to potential diversity of mycobiota recorded (Table 4). The ellipses mark the varieties colonized predominantly with xerospore (A) or gloeospore (B) producing fungi, resp.

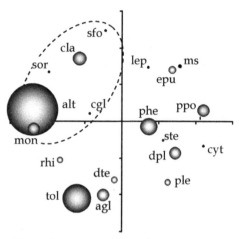

Fig. 14. Non-linear Map of fungal species as spectral variables. The codes of fungi are as given in Table 4. The size of balls is proportional to potential incidence of species recorded (Table 4). The ellipse marks the species causing fruit rot.

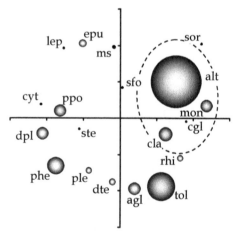

Fig. 15. Similarities in distribution of fungal species on stalk of sour cherry. The codes of fungi are as given in Table 4. The size of balls is proportional to potential incidence of species recorded (Table 4). The ellipse marks the species causing fruit rot.

The PCA of SC variables resulted eight components with few differences of their eigen-values. On our opinion this indicates the similar weight of numerous factors influencing on the incidence of fungal species in the orchard. On the NL map of PC variables of SPM the cluster of fruit rotting fungi became more compact. The grouping of the species remained unclear. All this underlines again the value of combined application of spectral mapping with PCA.

Neither sour cherry varieties nor associated fungi formed separated clusters on bivariate NLMaps (graphs are not shown). Most probably the two dimensional mapping distributed the information explained by PCs (see Tables 4 and 5) moderately thus the alterations shown by PC loadings were not reflected on scattering of PC scores (varieties or fungi) on the plot, i.e., the two dimensional NL-mapping smashed the variance and divided between two axes. One can frequently meet the clustering of variables on scatter plots and have no success in explanation. Nevertheless, this information is serviceable, because it might orient the further search and design of research project.

However, there were clear groups when tridimensional mapping carried out. In the case of sour cherry (Fig 16) neither the diversity nor the genetic background of varieties channelized the clustering.

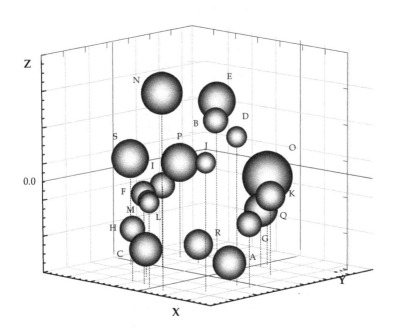

Fig. 16. Non-linear Map of sour cherry varieties as PC scores. The capital letters mark varieties as given in Table 4. The size of balls is proportional to diversity of colonization recorded (Table 4).

In the case of fungi the separation of species into two groups was more evident (Fig 17). Neither potential incidence nor mode of distribution is seemingly connected to clustering in tridimensional NL map. The first group comprises mainly fungi, which are pathogens or take part in fruit rot both on the tree and post harvesting in stores. Most of saprobionts are in the second groups. There was not found any relationships between taxonomic position of fungal species and their varietal dependent incidence in orchards. The high divergence of the composition of consortia on stalks might cause the absence of *Trichoderma* strains of biopreparation.

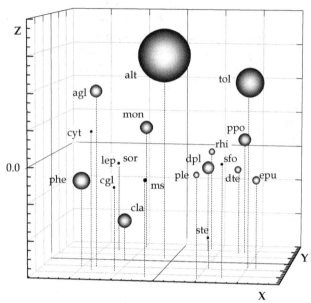

Fig. 17. Non-linear Map of fungal species as PC scores. The codes of fungi are as given in Table 4. The size of balls is proportional to potential incidence of species recorded (Table 3).

4. Conclusions

Presence of five *Glomerella* anamorphs has been established in Hungarian sour cherry orchards, among them *C. acutatum*, the new pathogen for Carpathian Basin became dominant over *C. gloeosporioides* in past five year. Using multivariate statistical methods for detailed analysis of mycobiota revealed two different fungal consortia associated to sour cherry in winter period.

PCA proved to be superior to multidimensional scaling and possible as well as valuable amendment of data exploration when combined with Potency and Spectral Mapping.

The correct order of organization of modules in multivariate analysis of data matrices comprising results of observations or experiments undergone by multifactorial effect can reveal unexpected factors thus increase the profitability of approach.

5. Acknowledgements

The research work was supported by the Baross Regional Innovation Programme (No. EA07-EA-KOZKFI-2008-0066). We thanks to Á. Szécsi for the critical reading of the manuscript.

6. References

Børve, J. & Stensvand, A. (2006). Timing of fungicide applications against anthracnose in sweet and sour cherry production in Norway. *Crop Protection*, 25(8), pp. 781-787. ISSN: 0261-2194

Børve, J., Djønne ,R. & Stensvand A. (2010). *Colletotrichum acutatum* occurs symptomatically on sweet cherry leaves. *European Journal of Plant Pathology*, 127(3), pp. 325-332. ISSN: 0929-1873

Frank, K.L., Kalkstein, L.S., Geils, B.W. & Thistle, H.W. (2008). Synoptic climatology of the long-distance dispersal of white pine blister rust. I. Development of an upper level synoptic classification. *International Journal of Biometeorology*, 52(7), pp.641-652. ISSN: 0020-7128

Frazer, G.W., Wulder, M.A. & Niemann, K.O. (2005). Simulation and quantification of the fine-scale spatial pattern and heterogeneity of forest canopy structure: A lacunarity-based method designed for analysis of continuous canopy heights. *Forest Ecology and Management*, 214(1-3), pp. 65-90. ISSN: 0378-1127

Garcia-Mozo, H., Gomez-Casero, M.T., Dominguez, E. & Galan, C. (2007). Influence of pollen emission and weather-related factors on variations in holm-oak (*Quercus ilex* subsp. *ballota*) acorn production. *Environmental and Experimental Botany*, 61(1), pp. 35-40. ISSN: 0098-8472

Jacquemin, B., Lanki, T., Yli-Tuomi, T., Vallius, M., Hoek, G., Heinrich, J., Timonen, K. & Pekkanen, J., (2009). Source category-specific PM(2.5) and urinary levels of Clara cell protein CC16. The ULTRA study. *Inhalation Toxicology*, 21(13), pp. 1068-1076. ISSN: 0895-8378

Lewi, PJ (2005). Spectral mapping, a personal and historical account of an adventure in multivariate data analysis. *Chemometrics and Intelligent Laboratory Systems*, 77(1-2), pp. 215-223. ISSN: 0169-7439

Lighthart, B., Shaffer, B.T., Frisch, A.S. & Paterno, D. (2009). Atmospheric culturable bacteria associated with meteorological conditions at a summer-time site in the mid-Willamette Valley, Oregon. *Aerobiologia*, 25(4), pp. 285-295. ISSN: 0393-5965

Magyar D (1998) Adatok a Budakeszi-erdQGmikroszkópikus gombáinak ismeretéhez. MSc, St. Steven University [Data to the Knowledge of the Microscopic Fungi in the Forests around Budakeszi]

Magyar D (2007) Aeromycological aspects of mycotechnology. In: *Mycotechnology: Current Trends and future Prospects*. (ed. M.K. Rai) I.K. International Publ. House, New Delhi: 226-263. ISBN: 9788189866082

Magyar, D., Eszeki, E.R., Oros, G., Szecsi, A., Kredics, L., Hatvani, L. & Körmöczi, P. (2011). The air spora of an orchid greenhouse. *Aerobiologia*, 27(2), pp. 121-134. ISSN: 0393-5965

Mardia, K.V., Kent, J.T. & Bibby, J.M. (1979). *Multivariate Analysis*, Academic Press, London, 213–254.

Mason E.W. (1937). *Annotated account of fungi received at the Imperial Mycological Institute*, List II, Fasc. 3., Kew, Surrey, IMI, pp . 69-99. ISSN 0027-5522 ; 3

Oros, G. & Naár, Z. (2008). Környezetkímélő biocidek és alkalmazásuk. HPO 0800405

Oros, G., Vajna, L., Balázs, K., Fekete, Z., Naár, Z. & Eszéki, E. (2010). Anthracnose and possibilities of the control with special regard to resident *Glomerella* population in sour cherry plantations of East Hungary. *Agricultural Research*, 39, pp. 12-17. ISSN 1587-1282

Oros, G., Naár, Z. & Cserháti, T. (2011). Growth response of *Trichoderma* species to organic solvents. *Molecular Informatics*, 30(2-3):276-285. ISSN: 1868-1743

Parrado, Z.G., Barrera, R.M.V., Rodriguez, C.R.F., Maray, A.M.V., Romero, R.P., Fraile, R. & Gonzalez, D.F. (2009). Alternative statistical methods for interpreting airborne Alder (*Alnus glutimosa* [L.] Gaertner) pollen concentrations. *International Journal of Biometeorology*, 53(1), pp. 1-9. ISSN: 0020-7128

Quraishi, T.A., Schauer, J.J. & Zhang, Y.X. (2009). Understanding sources of airborne water soluble metals in Lahore, Pakistan. *Kuwait Journal of Science & Engineering*, 36(1A), pp. 43-62. ISSN: 1024-8684

Pearson, K. (1901). On lines and planes of closest fit to systems of points in space. *Philosophical Magazine*, 6(2), pp. 559-572. ISSN: 1478-6435

Ribeiro, H., Cunha, M. & Abreu, I. (2007). Improving early-season estimates of olive production using airborne pollen multi-sampling sites RID A-6708-2010 RID A-6711-2010. *Aerobiologia*, 23(1), pp. 71-78. ISSN: 0393-5965

Ribeiro, H., Cunha, M. & Abreu, I. (2008). Quantitative forecasting of olive yield in Northern Portugal using a bioclimatic model RID A-6708-2010 RID A-6711-2010. *Aerobiologia*, 24(3), pp. 141-150. ISSN: 0393-5965

Rungratanaubon T., Panich, N. & Wangwongwattana, S. (2008). Characterization and source identification of trace metals inairborne particulates of Bangkok, Thailand. *Annals of the New York Academy of Sciences*, 1140, pp. 297-307. ISSN: 0077-8923

Sammon, J.W. (1969). A nonlinear mapping for data structure analysis, *IEEE Transactions on Computers*, 18, pp. 401-407. ISSN: 0018-9340

Schwarz, A., Wilcke, W., Kobza, J. & Zech, W. (1999). Spatial distribution of soil heavy metal concentrations as indicator of pollution sources at Mount Krizna (Great Fatra, central Slovakia) RID A-1573-2008. *Journal of Plant Nutrition and Soil Science*, 162(4), pp. 421-428. ISSN: 1436-8730

Shah, M.H., Shaheen, N., Jaffar, M., Khalique, A, Tariq, S.R. & Manzoor, S. (2006). Spatial variations in selected metal contents and particle size distribution in an urban and rural atmosphere of Islamabad, Pakistan. *Journal of Environmental Management*, 78(2), pp. 128-137. ISSN: 0301-4797

Shaheen, N., Shah, M.H. & Jaffar, M. (2005). A study of airborne selected metals and particle size distribution in relation to climatic variables and their source identification. *Water Air and Soil Pollution*, 164(1-4), pp.275-294. ISSN: 0049-6979

Titseesang, T., Wood, T. & Panich, N. (2008). Leaves of Orange Jasmine (*Murraya paniculata*) as Indicators of Airborne Heavy Metal in Bangkok, Thailand. *Annals of The New York Academy of Sciences*, 1140, pp. 282-289. ISSN: 0077-8923

Tokalioglu, S., Yilmaz, V. & Kartal, S. (2010). An Assessment on Metal Sources by Multivariate Analysis and Speciation of Metals in Soil Samples Using the BCR Sequential Extraction Procedure. *Clean-Soil Air Water*, 38(8), pp. 713-718. ISSN: 1863-0650

Uno, Y., Prasher, S.O., Lacroix, R., Goel, P.K., Karimi, Y., Viau, A. & Patel, R.M. (2005). Artificial neural networks to predict corn yield from Compact Airborne Spectrographic Imager data. *Computers and Electronics in Agriculture*, 47(2), pp. 149-161. ISSN: 0168-1699

Yang, C., Everitt, J.H. & Bradford, J.M. (2004). Airborne hyperspectral imagery and yield monitor data for estimating grain sorghum yield variability. *Transactions of The Asae*, 47(3), pp. 915-924. ISSN: 0001-2351

You, X.Q., Senthilselvan, A., Cherry, N.M., Kim, H.M. & Burstyn, I. (2008). Determinants of airborne concentrations of volatile organic compounds in rural areas of Western Canada. *Journal of Exposure Science and Environmental Epidemiology*, 18(2), pp. 117-128. ISSN: 1559-0631

Vasconcellos, P.C., Balasubramanian, R., Bruns, R.E., Sanchez-Ccoyllo, O. Andrade, M.F., Flues, M. (2007). Water-soluble ions and trace metals in airborne particles over urban areas of the state of Sauo Paulo, Brazil: Influences of local sources and long range transport RID C-2243-2011. *Water Air and Soil Pollution*, 186(1-4), pp. 63-73. ISSN: 0049-6979

8

PCA – A Powerful Method for Analyze Ecological Niches

Franc Janžekovič and Tone Novak
University of Maribor, Faculty of Natural Sciences and Mathematics,
Department of Biology, Maribor
Slovenia

1. Introduction

Principal Component Analysis, PCA, is a multivariate statistical technique that uses orthogonal transformation to convert a set of correlated variables into a set of orthogonal, uncorrelated axes called principal components (James & McCulloch 1990; Robertson et al., 2001; Legendre & Legendre 1998; Gotelli & Ellison 2004). Ecologists are most frequently dealing with multivariate datasets. This is especially true in field ecology, and this is why PCA is an attractive and frequently used method of data ordination in ecology. PCA enables condensation of data on a multivariate phenomenon into its main, representative features by projection of the data into a two-dimensional presentation. The two created resource axes are independent, and although they reduce the number of dimensions–i.e. the original data complexity–they maintain much of the original relationship between the variables: i.e., information or explained variance (Litvak & Hansell 1990). This is helpful in focusing attention on the main characteristics of the phenomenon under study. It is convenient that, if the first few principal components (PCs) explain a high percentage of variance, environmental variables that are not correlated with the first few PCs can be disregarded in the analysis (Toepfer et al., 1998). In addition, applying PCA has become relatively user-friendly because of the numerous programs that assist in carrying out the computational procedure with ease (Dolédec et al., 2000; Guisan & Zimmerman 2000; Robertson et al., 2001; Rissler & Apodaca 2007; Marmion et al., 2009).

PCA has been widely used in various fields of investigation and for different tasks. Many authors have used PCA for its main purpose: i.e., to reduce strongly correlated data groups or layers. These studies concern either environmental variation (e.g., Kelt et al., 1999; Johnson et al., 2006; Rissler & Apodaca 2007; Glor & Warren 2010; Novak et al., 2010a; Faucon et al., 2011; Grenouillet et al., 2011), the investigated species or community characteristics (e.g., Kingston et al., 2000; Pearman 2002; Youlatos 2004; Kitahara & Fujii 2005), or both, sometimes in combination with detrended correspondence analysis, DCA, canonical correspondence analysis, CCA, and other ordination methods (e.g., Warner et al., 2007; González-Cabello & Bellwood 2009; Marmion et al., 2009; Mezger & Pfeiffer 2011). The application of PCA has helped in various fields of ecological research, e.g., in determination of enterotypes of the human gut microbiome on the basis of specialization of their trophic niches (Arumugam et al. 2011). In aquatic habitat studies, it has been applied for evaluation of aquatic habitat suitability, its regionalization, analysis of fish abundance, their seasonal and spatial variation, lake ecosystem organization change etc. (Ahmadi-Nedushan et al., 2006; Blanck et al., 2007; Catalan et al., 2009). However, it has been often applied in analyzing farming system changes (Amanor & Pabi 2007).

In many cases, PCA has been used as a source or supporting analysis in the performance of more complex analysis, such as the study of adaptive fish radiation, strongly influenced by trophic niches and water depth (Clabaut et al., 2007), predicting the potential spatial extent of species invasion (Broennimann et al., 2007) and multi-trait analysis of intra- and interspecific variability of plant traits (Albert et al., 2010). Chamaillé et al. (2010) performed PCA and Hierarchical Ascendant Classification to evaluate environmental data, on the one hand, and human and dog population density data, on the other, in order to detect possible ranking of regions differently threatened by leishmaniasis.

Niche differentiation and partitioning is an ecological issue where PCA is frequently used. It enables efficient differentiation among related parapatric species (Dennis & Hellberg 2010). To access the problem authors use various available input data, which may be other than direct measurements of the niche. Since body shape and composition can readily be related to adaptation to the environment, morphometry figures as an adequate surrogate approach for studying the niche. Morphometric characteristics represent a data set vitable for evaluating the organism–environment relationship; besides PCA, Lecomte & Dodson (2005) additionally used discriminant analysis for this purpose. Inward et al. (2011) applied PCA to determine the morphological space of dung beetles representing regional faunas. Claude et al. (2003) demonstrated, using the case of turtles, that geometric morphometry, evaluated with PCA, can help to analyze the evolution of convergence. Morphometric differences between related species can easily refer to niche partitioning, reflecting differences in spatial or trophic level (Catalan et al. 2009; Niet-Castañda & Jiménez-Jiménez 2009; Novak et al., 2010b).

Hypogean habitats such as caves and artificial tunnels are relatively simple habitats owing to their low diversity, low production, and the constancy of their environmental factors (Culver 2005); they are thus suitable for investigating an environmental niche in situ. In this contribution we demonstrate the use of PCA in exploring an ecological niche in two case studies from caves in Slovenia. In the study of the three most abundant hymenopteran species that settled in the caves for rest, we only could evaluate their spatial niches on the basis of the usually measured environmental parameters. PCA was applied in two levels: 1. In the exploratory data analysis it was used as an efficacious tool to reduce the parameters into two principal components, PCs. 2. In the test hypothesis, the PCs of all three species were subjected to variance analysis to detect differences between the spatial niches of the three species.

1.1 Ecological niche concept

Ecological or environmental niche is one of the most useful concepts for exploring how and where organisms live and how are they related to their environment. After its introduction (Grinnel 1917), this concept changed considerably as new knowledge about the habitat and functioning of the organisms within was acquired. There are three main views in the evolution of the concept. The first view is that niche equals habitat, which is a multidimensional presentation of conditions in the local physical place where an organism lives. The second and most frequently applied understanding is the functioning of an organism within its concrete environment, which concerns acting on and responding to the organism's physical environment as well as to other organisms, originally, within a community. Since the community concept is in the course of radical change (Ricklefs 2008), it is convenient to replace the term community with a more general one, an assemblage. In practice, habitat and the function of an organism are often discussed as spatial, temporal and trophic niches. The third view is that the niche refers to variables within the whole range of the distribution area of an organism, which provides much information about organism–environment relationships on the global level (Soberón 2007).

The century-old niche concept has had many peaks and falls in the history. Since the initial idea was generated, it has evolved much over the years, being forgotten or neglected and/or misinterpreted, and recovering in different ways (Collwell & Rangel 2009). In the last decade, an intensive debate has taken place about redefining the meaning, importance and suitability of different aspects of the niche, including the measuring methods. This development accurately reflects the importance of its resurrection for the progress of ecological, evolutionary and related investigations. Today there are numerous niche concepts (Chase & Leibold 2003). Grinnell (1917) conceptualized the idea on a case study of a bird, the California Thrasher. He wrote that the niche comprehends the various circumstances to which a species is adapted by its constitution and way of living. He also wrote that no two species in a single fauna have precisely the same niche relationship, a fact which indicates the different roles of species within a community. What Grinnell called a niche was later understood for a long time as habitat, i.e., the sort of place where an organism lives. By a niche, Elton (1927) meant the place of an (animal) organism in its community, its relation to food and enemies and to some extant to other factors. He stressed the importance of the food (trophic) dimension of the ecological niche. Later (Elton 1933) he denoted a niche as a species mode of life, as for instance, professions in a human community. Instead of environments, Hutchinson (1957) attributed niches to species (Collwell & Rangel 2009), and explained the niche as part of an abstract multidimensional space, the ecospace, representing the whole range of intracommunity variables and interactions. Within this space, the way of life of a species is balanced by its tolerances and requirements. He called the overall potential niche the fundamental niche, in contrast to the realized niche, which is narrower for the negative impact of competitors and predators. The niche concept evolved additionally through discussion of several points of the competitive exclusion principle: i.e., the assumption that two species with identical environmental requirements cannot coexist indefinitely in the same location. According to Hardin (1960), the niche dimensions are represented by different abiotic and biotic variables concerning a species, such as its life history, habitat, position in the food chain, and its geographic range. Whittaker & Levin (1975) understood the niche as a species' requirements and its position in relation to other species in a given community. Recently, the importance of studying the niche to improve our understanding of the functioning of species and whole ecosystems has again become a widely discussed topic in global ecology. Pullian (2000) showed that, besides competition, other factors, such as niche width, habitat availability, dispersal, etc. influence the observed relationship between species distribution and the availability of suitable habitat, and should thus be incorporated into Hutchinson's niche concept. Soberón (2007) justified the separation of niches into two–the Grinnellian and the Eltonian class–on the basis of their focuses. Hutchinson's (1957) consideration can be applied to both groups. The Grinnellian class of niches is based on consideration of their non-interactive variables, such as average temperature, precipitation and solar radiation, and environmental conditions on a broad scale. These variables are relevant to understanding coarse-scale ecological and geographic properties of species. The Eltonian class niches, in contrast, focus on bionomic variables, such as biotic interactions and resource–consumer dynamics, which can be measured principally on a local scale. Whereas datasets of variables of the Grinnellian niche group have been rapidly compiled in the World, very little theory has been developed explicitly about this. On the other hand, variables for considering much more dynamic and complex Eltonian niches have never been available (Soberón 2007). Both classes of niches are relevant to understanding the distribution of individuals of a species, but the Eltonian class is easier to measure at the high spatial resolutions characteristic of most ecological studies, whereas the Grinnellian class is suited to the low spatial resolution at which distributions are typically defined (Soberón 2007). Applying the modelling of species distribution to the distribution constraints is strongly encouraged to provide better insight in species distributions (Kearney & Porter 2009; Bellier et al., 2010). It is important to understand that a niche is not a conservative concept, but a consequence of the complexity of the subject, which may refer to very different features of the fundamental niche, with different ecological and evolutionary properties (Soberón & Nakamura 2009). It has been demonstrated that, on the one hand, inconsistent adaptive pressures may give rise to a whole palette of niche diversification (e.g., Romero 2011), while, on the other hand, convergent evolution in various combinations takes place within the multidimensional niche space (e.g., Hormon et al., 2005).

1.2 Ordination and the PCA concept

Ordination is a method in multivariate analysis used in exploratory data analysis. Exploratory data analysis is an approach to analyzing data sets to summarize their main characteristics in an easy-to-understand form, often in graphs. In this procedure no statistical modelling is used. The order of objects in ordinations is characterized by values of multiple variables. Similar objects are ordinated near each other and vice versa. Many ordination techniques exist, including principal components analysis (PCA), non-metric multidimensional scaling (NMDS), correspondence analysis (CA) and its derivatives, like detrended CA (DCA), canonical CA (CCA), Bray–Curtis ordination, and redundancy analysis (RDA), among others (Legendre & Legendre 1998; Gotelli & Ellison 2004).

PCA is widely useful in considering species; it is appropriate for the analysis of community composition data or as gradient analysis. Gradient analysis is an analytical method used in plant community ecology to relate the abundance of various species within a plant community to various environmental gradients by ordination or by weighted averaging. These gradients are usually important in plant species distribution, and include temperature, water availability, light, and soil nutrients, or their closely correlated surrogates (Lepš & Šmilauer 2003).

2.1 Environmental niche of three hymenopteran and two spider species

Between 1977 and 2004, 63 caves and artificial tunnels were ecologically investigated in Slovenia; the three most abundant Hymenoptera species found in these studies have been ecologically evaluated (details in Novak et al. 2010a). In the caves, many environmental data were collected, as follows. The following abbreviations of the environmental variables are used: Dist-E = distance from entrance; Dist-S = distance from surface; Illum = illumination; PCS = passage cross-section; Tair =air temperature; RH = relative air humidity; Tgr = ground temperature; HY = substrate moisture. The hymenopteran spatial niche breadth was originally represented by nine variables. The variation was subjected to PCA, and differences in niche overlap were tested using One-way ANOVA. In the following, we demonstrate the analysis of occupied physical space in the three species: *Amblyteles armatorius*, n=16, *Diphyus quadripunctorius*, n=42, and *Exallonyx longicornis*, n=44. These variables refer to the environmental conditions for the individual placements within the caves.

PCA requires normal data distribution. This is often not the case with the environmental data provided by field investigations, as in our case. In variables presented as proportions or ratios, e.g., humidity, this problem can be overcome with the arc-sin transformation. In those variables stretched over a large scale of values, e.g., illumination and passage cross section, this can be achieved by transformation in the logarithmic scale. In our study, we used the Kolmogorov-Smirnov test, K-S, to check the data for normality. To normalize distribution, we transformed air humidity and substrate moisture data (*arcsin*) (Fig. 1), and passage cross section and illumination data (*log*) (Fig. 2). PCA is sensitive to the relative scaling of the original variables. We therefore z-standardized the data. Here we demonstrate relations between nine environmental variables with Pearson correlation coefficients (Table 1).

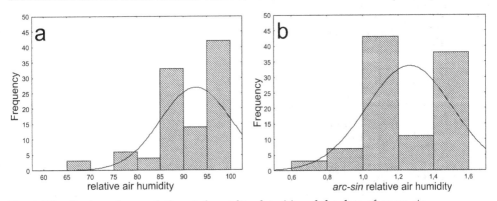

Fig. 1. Distribution of row relative air humidity data (a) and the data after *arc-sin* transformation (b) with normal distribution curve.

To obtain detailed information on the pattern of variation, the sets of nine environmental variables were subjected to PCA. In this way, we obtained nine PCs. These new values are called principal component scores. The Eigenvalue and ratios of explained variances are presented in Table 2, where PC variance is in progressive decline. The last four components represent such a small ratio of the total variance that it is reasonable to ask whether they describe any biotic response or not. A common rule is to interpret only those components that contribute more than 5% of the total variance. In this study case on Hymenoptera, PCs1 to PCs5 meet this criterion in the total account of 92.5% of the variance explained, while 7.5% of the variance remains unexplained. The explained contribution of variances to the total variance is shown in a scree plot (Fig. 3). The large differences between the variances of the first three PCs and much smaller ones of the other scores are clearly evident.

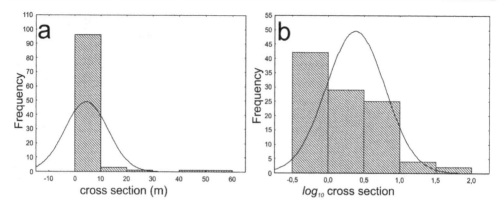

Fig. 2. Distribution of row cross section data (a) and the data after logarithmic transformation (b) with normal distribution curve.

	1	2	3	4	5	6	7	8	9
1 Air temperature	1.00 ---								
2 *arc-sin* relative air humidity	0.15 0.133	1.00 ---							
3 Ground temperature	**0.94** **<0.001**	0.18 0.079	1.00 ---						
4 *arc-sin* substrate moisture	**0.388** **<0.001**	**0.59** **<0.001**	**0.37** **<0.001**	1.00 ---					
5 Airflow	**-0.48** **<0.001**	**-0.36** **<0.001**	**-0.43** **<0.001**	**-0.55** **<0.001**	1.00 ---				
6 Distance from entrance	**-0.34** **<0.001**	0.14 0.153	**-0.41** **<0.001**	0.10 0.312	0.04 0.712	1.00 ---			
7 Distance from surface	-0.02 0.837	**0.24** **0.017**	-0.04 0.683	**0.46** **<0.001**	-0.11 0.275	**0.67** **<0.001**	1.00 ---		
8 Passage cross-section	**0.35** **<0.001**	0.17 0.089	**0.23** **0.025**	**0.39** **<0.001**	**-0.40** **<0.001**	-0.11 0.274	0.05 0.656	1.00 ---	
9 *log* illumination	**0.45** **<0.001**	-0.18 0.077	**0.46** **<0.001**	-0.04 0.690	-0.07 0.494	**-0.821** **<0.001**	**-0.679** **<0.001**	**0.37** **<0.001**	1.00 ---

Table 1. Pearson correlations coefficient among nine environmental variables. Significant correlations in bold. (Upper row r, lower row p).

PC	Eigenvalue	% Total	Cumulative Eigenvalue	Cumulative %
1	3.38	37.61	3.38	37.61
2	2.65	29.48	6.04	67.09
3	0.94	10.44	6.98	77.52
4	0.80	8.94	7.78	86.47
5	0.54	6.05	8.33	92.52
6	0.37	4.06	8.69	96.58
7	0.18	1.95	8.87	98.53
8	0.09	1.03	8.96	99.55
9	0.04	0.45	9.00	100.00

Table 2. Eigenvalues and percentages of explained variability.

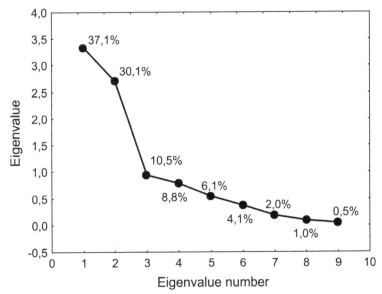

Fig. 3. Scree plot of the eigenvalue and the percent of variance explained by each component is shown in decreasing order.

Projection of the variables on the factor plane revealed that the 1st and the 2nd axes of the PCs explained 37.6% and 29.5% of the total variance. The Pearson correlation coefficients and elementary graphics associated with the relations between PCs and environmental variables are presented in Table 3 and Fig 4, respectively. In this graphic presentation, they are placed on a circle, called the correlation circle, with the pair of factor axes as its axes. The stronger the correlation between a variable and the factor, the greater the correlation of the corresponding variable with the factor axes. The variables that are correlated with a particular factor can thus be identified, thereby providing information as to which variables can explain the given factor. This is demonstrated in Fig.5. PC1 best explains the variability of air and ground temperature, and illumination: these values increase with the decreasing PC1, while the values of airflow and distance from the entrance increase with the increasing

PC1. PC2 best explains the variability of air humidity, substrate moisture and both distance from the entrance and from the surface: these values increase with the increasing PC2. In the case presented the explanatory power of PCA with respect to variable importance is evident. PCs thus fully represent adequate surrogates to explain the spatial component of the niche. For the interpretation of these outputs, one needs good biological and ecological knowledge about the organisms under study. The projections of the environmental dimensions of the three species are represented by polygons in Fig. 5. A more elaborate figure has been published elsewhere (Novak et al. 2010a).

Parameter \ PC	PC1	PC2
Air temperature	-0.87	-0.10
arc-sin relative air humidity	-0.29	-0.60
Ground temperature	-0.85	-0.06
arc-sin substrate moisture	-0.49	-0.72
Airflow	0.62	0.45
Distance from entrance	0.60	-0.67
Distance from surface	0.26	-0.82
log passage cross-section	-0.55	-0.20
log illumination	-0.69	0.63
$F_{2,99}$; p	19.85; <0.001	9.19; <0.001

Table 3. Pearson correlation coefficient between environmental variables and the first two Principal Components (PCs) F and p values of one-way ANOVA in testing the first two PCs according to the three hymenopteran species.

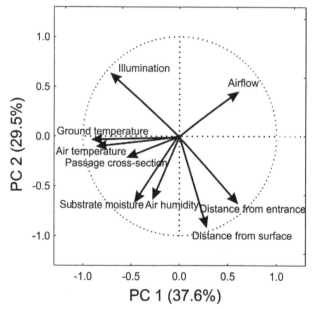

Fig. 4. Projection of the nine ecological variables on the 1st and 2nd factor planes. Graphical associated fall of the variables (arrows) in the correlation circle.

Moreover, PCs enable the testing of differences between environmental niches. For this purpose, in the test hypothesis, the PCs defining niches were subjected to variance analysis for differences between the three species. One-way ANOVA was used to test differences between species in the 1st and 2nd principal components (F and p values in Table 3). In this way, PCA allows testing of differences between niches.

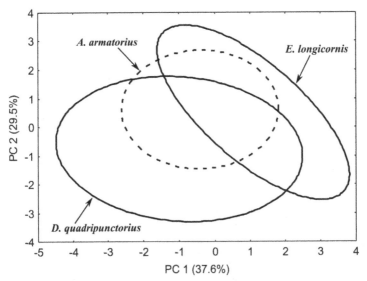

Fig. 5. Ordination of the nine environmental variables in 1st and 2nd PC axes. Ellipses (95% confidence) represent spatial niches in the three hymenopteran species.

The same analyses of the spatial niches were carried out on two co-existing spider species, *Meta menardi* and *Metellina meriannae* (Novak et al. 2010b). In this case, the variations in temperature, humidity, airflow and illumination were subjected to PCA. The 1st and the 2nd PCs together explained 70.4% of variation (Figs. 6 and 7). In this way, we presented the course of temporal changes in the spatial niches of the two spiders.

3. Discussion

Since computer techniques and technologies have enabled efficacious computation of PCA, it has become one of the most useful tools in ecology in various fields of use. Still, one can readily notice that many problems appear when its applicability for different purposes is to be estimated. On the one hand, reservations occur because of the credibility or interpretability of the data. Yet Austin (1985), e.g., stated that animal ecologists often use PCA without discussion of the ecological implications of its linear model, although the PCA axes are not necessarily ecologically independent, and there is no necessary ecological interpretation of components. Besides, it is particularly notable that two- and three-dimensional data using Gauss species response curves can produce complex flask-like distortions in which the underlying gradient structure is impossible to recognize without prior knowledge. In this sense, some authors (e.g., Hendrickx et al., 2007), in a specific context, decided not to rely on the obtained PCA axes, since they obscured additive and

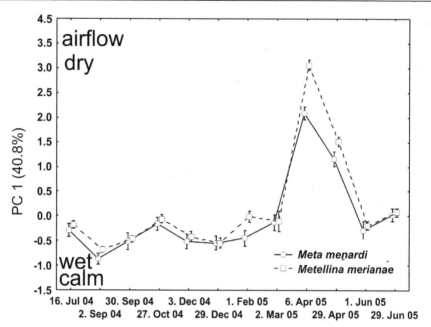

Fig. 6. Comparison between temporal spatial niches of *Meta menardi* and *Metellina merianae* in their 1st principal component (mean ± SE).

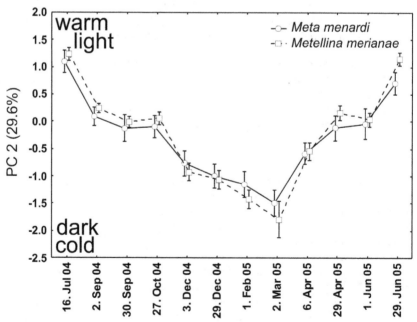

Fig. 7. Comparison between temporal spatial niches of *Meta menardi* and *Metellina merianae* in their 2nd principal component (mean ± SE).

interactive effects among variables that were partially correlated. The others highlighted that the use of PCA on spatially autocorrelated data is not appropriate in a spatial context (Novembre & Stephens 2008). Indirectly, this is the identical aspect of the problem that Nakagawa & Cuthill (2007) explain on a general level; they direct attention to the mostly neglected fact that we are still dealing with and discussing statistical rather than biotic importance. Null hypothesis significance testing, e.g., does not provide us with the magnitude of an effect of interest, nor with the precision of the estimate of the magnitude of that effect. They thus advocate (ibid.) presentation of biotic relevance: i.e., measurement of the magnitude of an effect on an organism and its confidence interval. These are the real goals in biology irrespective of the statistical values measured. In this sense, biologists and ecologists are often "trapped in statistics", into "statitraps", rather than dealing with biotic phenomena themselves. In addition, in comparison with other methods, PCA sometimes proves not to be as efficacious. In their study on ecological niches of two *Ceratitis* flies, De Meyer et al. (2008) aver that the PCA model is apparently not as good as the GARP (genetic algorithm for rule-set prediction) model at capturing the species–environment relationship. This is probably because the PCA model cannot account for nonlinear species–environment relationships in the way that GARP can.

Additionally, Ricklefs (2008) established that a local ecological community, which consists of those species whose distribution include a particular point in space and time, is an epiphenomenon with relatively little explanatory power in ecology and evolutionary biology. To understand the coexistence of species locally, one must understand what shapes species distribution within regions, but factors that constrain distribution within regions are poorly understood. This evidence of the disintegrating concept of "community" requires reconsideration of our prior explanations of coexisting species in the subterranean environment. Although many authors use the term community when dealing with its biota, their methods of discussion reveal that they mostly use the term in its wide, much looser meaning rather than referring to specific species composition and their inter- and intraspecific relations. This is especially evident in their continual references to the fact that the biology and/or ecology of many species remain unknown.

In this contribution we had presented common uses of PCA and its efficacy in accessing multivariate data, such as provided in the most usual and convenient field investigations in caves. With respect to the dynamic niche conception (Soberón 2007; Soberón & Nakamura 2009), our case studies deserve additional comment. On the one hand, the essence of the "disintegration community concept" sensu Ricklefs (2008) can well be perceived in the widely distributed species that we encounter in caves. These include the animal species appertaining to the parietal association, i.e., those species found especially on the walls and ceiling near cave entrances (Jennings 1997). The three hymenopteran and two spider species discussed in this study belong among them. The parietal association is a heterogeneous assemblage of species with respect to their geographical distribution and activity in the subterranean environment, thus directly raising the need for more complete knowledge of their "biogeography of the species" sensu Ricklefs (2008). On the other hand, in highly endemic troglomorphic species–i.e., those well adapted to the hypogean habitat (e.g., Gibert and Deharveng 2002, Christman et al., 2005, Culver and Pipan 2009)–the community disconception might have little or no significant impact on the findings under discussion.

4. Conclusions

PCA is a useful tool enabling ordination of environmental variables in ecology. An environmental niche comprises multidimensional data set, while PCA is an appropriate statistical tool to handle such data sets. In this way, PCA readily provides the means to explain the variance magnitudes related to environmental variables, which represent the environmental niche. Despite that, one must be aware that PCA output depends on input data, which can never cover all dimensions of an environmental niche. Besides, PCA may obscure many effects among partially correlated variables. As with other statistical approaches, it is necessary to consider the results carefully, implementing a broad knowledge of the biology and ecology of the organisms under study in order to avoid statistical artifacts.

5. Acknowledgement

We are indebted to Michelle Gadpaille for valuable improvement of the language. This study was partly supported by the Slovene Ministry of Higher Education, Science and Technology within the research program Biodiversity (grant P1-0078).

6. References

Ahmadi-Nedushan, B., St-Hilaire, A., Bérubé, M., Robichaud, E., Thiémonge, N. & Bobée, B. (2006). A review of statistical methods for the evaluation of aquatic habitat suitability for instream flow assessment. *River Research and Applications*, Vol. 22, pp. 503-523.

Albert, C.H., Thuiller, W., Yoccoz, N.G., Douzet, R., Aubert, S. & Lavorel, S. (2010). A multi-trait approach reveals the structure and the relative importance of intra- vs. interspecific variability in plant traits. *Functional Ecology*, Vol. 24, pp. 1192-1201.

Amanor. K.S. & Pabi, O. (2007). Space, time, rhetoric and agricultural change in the transition zone of Ghana. *Human Ecology*, Vol. 35, pp. 51−67.

Arumugam, M., Raes, J., Pelletier, E. *et al.* (2011). Enterotypes of the human gut microbiome. *Nature*, Vol. 473, pp. 174-180.

Austin, M.P. (1985). Continuum concept, ordination methods, and niche theory. *Annual Review of Ecology and Systematics*, Vol. 16, pp. 39−61.

Bellier E., Certain G., Planque B., Monestiez P. & Bretagnolle, V. (2010). Modelling habitat selection at multiple scales with multivariate geostatistics: an application to seabirds in open sea. *Oikos*, Vol. 119, pp. 988-999.

Blanck, A., Tedesco, P.A. & Lamouroux, N. (2007). Relationships between life-history strategies of European freshwater fish species and their habitat preferences. *Freshwater Biology*, Vol. 52, pp. 843-859.

Broennimann, O., Treier, U.A., Müller-Schärer, H., Thuiller, W., Peterson, A.T. & Giusan, A. (2007). Evidence of climatic niche shift during biological invasion. *Ecology Letters*, Vol. 10, pp. 701-709.

Catalan, J., Barbieri, M.G., Bartumeus, F., Bitušík, P., Botev., I., Brancelj, A., Cogălcineanu, D. et al. (2009). Ecological thresholds in European alpine lakes. *Freshwater Biology*, Vol. 54, pp. 2494-2517.

Chamaillé. L., Tran, A., Meunier, A., Bourdoiseau, G., Ready, P. & Dedet, J.-P. (2010). Environmental risk mapping of canine leishmaniasis in France. *Parasites & Vectors*, Vol. 3, No.31, pp. 8.

Chase, J.M. & Leibold, M. (2003). *Ecological Niches: Linking Classical and Contemporary Approaches.* University of Chicago Press.

Christman, M.C., Culver, D.C., Madden, M. & White, D. (2005). Patterns of endemism of the eastern North American cave fauna. *Journal of Biogeography*, Vol. 32, pp. 1441–1452.

Clabaut, C., Bunje, P.M.E., Salzburger, W. & Meyer, A. (2006). Geometric morphometrical analyses provide evidence for the adaptive character of the Tanganyikan cichlid fish radiation. *Evolution*, Vol. 61, pp. 560–578.

Claude, J., Paradis, E., Tong, H., Auffray, J.-C. (2003). A geometric morphometric assessment of the effects of environment and cladogenesis on the evolution of the turtle shell. *Biological Journal of the Linnean Society*, Vol. 79, pp. 485–501.

Collwell, R.K. & Rangel, T.F. (2009). Hutchinson's duality: The once and future niche. *Proceedings of the National Academy of Sciences of the United States of America*, Vol. 106, Suppl. 2, pp. 19651–19658.

Culver, D.C. (2005). Ecotones. In: Culver, D., White, W.B. (Eds.), *Encyclopedia of Caves*. Elsevier, Amsterdam, pp. 206–208.

Culver, D.C. & Pipan, T. (2009). *The Biology of Caves and Other Subterranean Habitats*. Oxford University Press.

De Meyer, M., Robertson, M.P., Peterson, A.T. & Mansell, M.W. (2008). Ecological niches and potential geographical distributions of Mediterranean fruit fly (*Ceratitis capitata*) and Natal fruit fly (*Ceratitis rosa*). *Journal of Biogeography*, Vol. 35, pp. 270–281.

Dennis, A.B. & Hellberg, M.E. (2010). Ecological partitioning among parapatric cryptic species. *Molecular Ecology*, Vol. 19, pp. 3206–3225.

Dolédec, S., Chessel, D. & Gimaret-Carpentier, C. (2000). Niche Separation in Community Analysis: A New Method. *Ecology*, Vol. 81, pp. 2914-2927.

Eaton, M.D., Soberón, J. & Townsend Petersen, A. (2008). Phylogenetic perspective on ecological niche evolution in American blackbirds (Family Icteridae). *Biological Journal of the Linnean Society*, Vol. 94, pp. 869–878.

Ecke, F., Löfgren, O., Hörnfeldt, B., Eklund, U, Ericsson, P. & Sörlin, D. (2001). Abundance and diversity of small mammals in relation to structural habitat factors. *Ecological Bulletins*, Vol. 49, pp. 165–171.

Elton C. (1927). *Animal Ecology.* Sidgwick & Jackson. London.

Elton, C. (1933) *The Ecology of Animals*. Methuen, London.

Faucon, M.P., Parmentier, I., Colinet, G., Mahy, G., Luhembwe, M.N. & Meerts, P. (2011). May rare metallophytes benefit from disturbed soils following mining activity? The Case of the *Crepidorhopalon tenuis* in Katanga (D. R. Congo). *Restoration Ecology*, Vol. 19, pp. 333-343.

Gibert, J. & Deharveng, L. (2002). Subterranean ecosystems: a truncated functional biodiversity. *Bioscience*, Vol. 52, pp. 473–481.

Glor, R.E. & Warren, D. (2010). Testing ecological explanations for biogeographic boundaries. *Evolution*, Vol. 65, pp. 673-683.

González-Cabello, A. & Bellwood, D.R. (2009). Local ecological impacts of regional biodiversity on reef fish assemblages. *Journal of Biogeography*, Vol. 36, pp. 1129-1137.

Gordon, W., Frazer, G.W., Wulder, M.A. & Niemann, K.O. (2005). Simulation and quantification of the fine-scale spatial pattern and heterogeneity of forest canopy structure: A lacunarity-based method designed for analysis of continuous canopy heights. *Forest Ecology and Management*, Vol. 3, pp. 65-90.

Gotelli, N.J. & Ellison, A.M. (2004). *A Primer of Ecological Statistics*. Sinauer Associates, Inc. Sunderland. 510 pp.

Grenouillet, G., Buisson, L., Casajus, N. & Lek, S. (2011). Ensemble modelling of species distribution: the effects of geographical and environmental ranges. *Ecography*, Vol. 34, pp. 9-17.

Grinnell, J. (1917). The niche-relationships of the California Thrasher. *Auk*, Vol. 34, pp. 427-433.

Guisan, A. & Zimmerman, N.E. (2000). Predictive habitat distribution models in ecology. *Ecological Modelling*, Vol. 135, pp. 147-186.

Hardin, G. (1960). The competitive exclusion principle. *Science*, Vol. 131, pp. 1292-1297.

Hendrickx, F., Maelfait, J.-P., Van Wingerden, W., Schweiger, O., Speelmans, M., Aviron, S., Augenstein, I., et al. (2007). How landscape structure, land-use intensity and habitat diversity affect components of total arthropod diversity in agricultural landscapes. *Journal of Applied Ecology*, Vol. 44, pp. 340-351.

Hormon, L.J., Kolbe, J.J., Cheverud, J.M. & Losos, J.B. (2005). Convergence and the multidimensional niche. *Evolution*, Vol. 59, pp. 409-421.

Hutchinson, G.E. (1957). *Concluding remarks*. In: *Cold Spring Harbor Symposia on Quantitative Biology*, Vol. 22, pp. 415-427.

Inward, D.J.G., Davies, R.G., Pergande, C., Denham, A.J. & Vogler, A.P. (2011). Local and regional ecological morphology of dung beetle assemblages across four biogeographic regions. *Journal of Biogeography*, Vol. 38, pp. 15.

James, F.C. & McCulloch, C.E. (1990). Multivariate Analysis in Ecology and Systematics: Panacea or Pandora's Box? *Annual Review of Ecology and Systematics*, Vol. 21, pp. 129-166.

Jennings, J.N. (1997). *Cave and karst terminology*. Coggan M, Nicholson P, eds. Australian Speleological Federation Incorporated Administrative Handbook. http://home.mira.net/~gnb/caving/papers/jj-cakt.html

Johnson, R.K., Hering, D., Furse, M.T. & Clarke, R.T. (2006). Detection of ecological change using multiple organism groups: metrics and uncertainty. In: Furse, M.T., Hering, D., Brabec, K., Buffagni, A., Sandin L. & Verdonschot P.F.M. (eds), The Ecological Status of European Rivers: Evaluation and Intercalibration of Assessment Methods. *Hydrobiologia*, Vol. 566, pp. 115-137.

Kearney, M. & Porter, W. (2009). Mechanistic niche modelling: combining physiological and spatial data to predict species' ranges. *Ecology Letters*, Vol. 12, pp. 334-350.

Kelt, D.A., Meserve, P.L., Bruce, Patterson, D. & Lang, B.K. (1999). Scale Dependence and Scale Independence in Habitat Associations of Small Mammals in Southern Temperate Rainforest. *Oikos*, Vol. 85, pp. 320-334.

Kingston, T., Jones, G., Zubaid, A. & Kunz, T.H. (2000). Resource partitioning in Rhinolophoid bats revisited. *Oecologia*, Vol. 124, pp. 332-342.

Kitahara, M. & Fujii, K. (2005). Analysis and understanding of butterfly community composition based on multivariate approaches and the concept of generalist/specialist strategies. *Entomological Science*, Vol. 8., pp. 137–149.

Lecomte, F. & Dodson, J.J. (2005). Distinguishing trophic and habitat partitioning among sympatric populations of the estuarine fish *Osmerus mordax* Mitchill. *Journal of Fish Biology*, Vol. 66, 1601–1623.

Legendre, P., & Legendre, L., (1998). Numerical Ecology. Elsevier: Amsterdam, 853 pp.

Lepš, J. & Šmilauer, P. (2003). *Multivariate analysis of ecological data using CANOCO*. Cambridge University Press. Cambridge.

Litvak, M.K. & Hansell, R.I.C. (1990). A community perspective on the multidimensional niche. *Journal of Animal Ecology*, Vol. 59, pp. 931–940.

Marmion, M., Parviainen, M., Luoto, M., Heikkinen, R.K. & Thuiller, W. (2009). Evaluation of consensus methods in predictive species distribution modelling. *Diversity and Distributions*, Vol. 15, pp. 59–69.

Mezger, D. & Pfeiffer, M. (2011). Partitioning the impact of abiotic factors and spatial patterns on species richness and community structure of ground ant assemblages in four Bornean rainforests. *Ecography*, Vol. 34, pp. 39–48.

Nieto-Castañeda, I.G. & Jiménez-Jiménez, M.L. (2009). Possible niche differentiation of two desert wandering spiders of the genus *Syspira* (Araneae: Miturgidae). *The Journal of Arachnology*, Vol. 37, pp. 299–305.

Nakagawa, S. & Cuthill, I.C. (2007). Effect size, confidence interval and statistical significance: a practical guide for biologists. *Biological Reviews*, Vol. 82, pp. 591–605.

Novak, T. (2005). Terrestrial fauna from cavities in Northern and Central Slovenia, and a review of systematically ecologically investigated cavities. *Acta Carsologica*, Vol. 34, 169–210.

Novak, T., Thirion, C., Janžekovič, F. (2010a). Hypogean ecophase of three hymenopteran species in Central European caves. *Italian Journal of Zoology*, Vol. 77, pp. 469–475.

Novak, T., Tkavc, T., Kuntner, M., Arnett, A.E., Lipovšek Delakorda, S., Perc, M. & Janžekovič, F. (2010b). Niche partitioning in orbweaving spiders *Meta menardi* and *Metellina merianae* (Tetragnathidae). *Acta Oecologica*, Vol. 36, pp. 522–529.

Novembre, J. & Stephens, M. (2008). Interpreting principal component analysis of spatial population genetic variation. *Nature Genetics*, Vol. 40, pp. 646–649.

Pearman, P.B. (2002). The scale of community structure: Habitat variation and avian guilds in tropical forest understory. *Ecological Monographs*, Vol. 72, pp. 19–39.

Pulliam, H.R. (2000). On the relationship between niche and distribution. *Ecology Letters*, Vol. 3, pp. 349–361.

Ricklefs, R. (2008). Disintegration of the ecological community. *The American Naturalist*, Vol. 172, pp. 741–750.

Rissler, L.J. & Apodaca, J.J. (2007). Adding more ecology into species delimitation: Ecological niche models and phylogeography help define cryptic species in the black salamander (*Aneides flavipunctatus*). *Systematic Biology*, Vol. 56, pp. 924–942.

Robertson, M.P., Caithness, N. & Villet, M.H. (2001). A PCA-based modelling technique for predicting environmental suitability for organisms from presence records. *Diversity and Distributions*, Vol. 7, pp. 15–27.

Romero, A. (2011). The evolution of cave life. *American Scientist*, Vol. 99, pp. 144–151.

Soberón, J. (2007). Grinnelian and Eltonian niches and geographic distributions of species. *Ecology Letters*, Vol. 10, pp. 1115–1123.

Soberón, J. & Nakamura, M. (2009). Niches and distributional areas: Concepts, methods, and assumptions. *Proceedings of the National Academy of Sciences of the United States of America*, Vol. 106, Suppl. 2, pp. 19644–19650.

Toepfer, C.S., Williams, L.R., Martinez, A.D. & Fisher, W.L. (1998). Fish and habitat heterogeneity in four streams in the central Oklahoma/Texas plains ecoregion. *Proceedings of the Oklahoma Academy of Science* 78: 41–48.

Warner, B.G., Asada, T. & Quinn, N.P. (2007). Seasonal influences on the ecology of testate amoebae (Protozoa) in a small *Sphagnum* peatland in Southern Ontario, Canada. *Microbial Ecology*, Vol. 54, pp. 91–100.

Whittaker, R.H. & Levin S.A., eds. (1975). *Niche*. Benchmark Papers in Ecology. Dowden, Hutchinson & Ross Inc., Stroudsburg.

Youlatos, D. (2004). Multivariate analysis of organismal and habitat parameters in two neotropical primate communities. *American Journal of Physical Anthropology*, Vol. 123, pp. 181–194.

9

Principal Component Analysis Applied to SPECT and PET Data of Dementia Patients – A Review

Elisabeth Stühler and Dorit Merhof
University of Konstanz
Germany

1. Introduction

Alzheimer's disease (AD) is the most common cause of dementia, followed by vascular and frontotemporal dementia. Approximatly 8% of the population in developed countries is impaired by AD at the age of 65, with the risk expanding to 30% for individuals over the age of 85 years (Petrella et al. (2003)). Due to the increasing life expectancy, the spread of AD is estimated to triple over the next 50 years (Petrella et al. (2003)). If AD remained untreated, the economic impact on society would increase dramatically (Carr et al. (1997); Mueller et al. (2005)), but it is even more important to alleviate the psychological strain on patients and their relatives. Normally, a patient affected by AD has an anticipated average life expectancy of 8-10 more years, divided into several stages of the disease. The neuropathological stages of AD are described in detail in Braak & Braak (1991), where the development of amyloid deposition and neurofibrillary changes within the brain are explained. These changes can already be observed in the preclinical phase, i.e., before clinical symptoms occur. Clinical symptoms usually begin (in early stages) with memory and learning impairment, followed by alterations in judgement, display of social behavioral problems and reduced faculty of speech. In late stages of AD, motoric and sensory functions are affected as well (Selkoe (2001)).

First pharmaceuticals for treatment of AD symptoms were recently developed, and there are several more under clinical trials at the moment, which in turn require the early detection of AD (Petrella et al. (2003)). Cases with early-onset AD are usually diagnosed with mild cognitive impairment (MCI). According to Tabert et al. (2006), about 10% of cases with amnestic MCI (i.e., patients with memory deficits) and about 50% of MCI cases with further cognitive domain deficits will convert to AD within three years.

In early stages of AD, structural changes within the brain are difficult to detect, as they are restrained to very specific areas (e.g., hippocampal atrophy) until AD is advanced to a middle or later stage. Petrella et al. (2003) advise therefore to resort to nuclear medicine imaging which captures more subtle pathological changes, rather than to magnetic resonance imaging (MRI) or X-ray computed tomography (CT) as they are less capable for early detection of dementia. Prevalent in clinical assessment of AD are positron emission tomography (PET) and single-photon emission computed tomography (SPECT), where PET is observed to perform superior to SPECT for distinguishing between AD and a control group (CTR), e.g., in Herholz et al. (2002b).

In nuclear medicine, the biomarkers used for detection of AD include increased β-amyloid deposition, decreased glucose metabolism and reduced blood flow in the brain, which are among many indicators for AD. Furthermore, AD can be correlated to several risk factors, such as the genetic inheritance of the $\epsilon 4$ allele of the apolipoprotein E (APOE) or the increased accumulation of tau proteins in the cerebrospinal fluid (CSF).

SPECT or PET images are typically evaluated by clinical reading, but this procedure requires expert knowledge, is time-consuming and rater-dependent. Therefore, statistical analyses for automated detection or prediction of AD progression in MCI have been subject to recent research.

As SPECT or PET datasets contain a large amount of information, i.e., more than 10^5 voxel-values within the whole-brain region, and as usually up to 100 subjects are considered in a study, statistical analysis of the 3D-images is very challenging. It includes univariate analysis where a voxel-wise comparison is performed to differentiate between AD and normal controls (CTR), e.g., in Dukart et al. (2010) and Habeck et al. (2008). More recently also multivariate analysis, such as principal component analysis (PCA), has been applied to enable statistical evaluation of all voxel-values at the same time, thereby accounting not only for differences in single intensity values but also correlations between regions. This usually outperforms univariate analysis in the early identification of AD (Habeck et al. (2008)).

In many studies, PCA is therefore employed to either reduce the high dimensionality of the data (Markiewicz et al. (2009; 2011a;b); Merhof et al. (2009; 2011)), to discriminate between dementia of Alzheimer type and asymptomatic controls (Fripp et al. (2008a); Habeck et al. (2008); Scarmeas et al. (2004)) or to assess the amount of variability of the data (Fripp et al. (2008a;b)).

The objective of this review is to present and discuss these applications of PCA, and also to give an insight into adequate preprocessing of the data and implementation of PCA:

Basically, any analysis of PET or SPECT data requires preprocessing of the data in a first step, comprising registration of each subject to a brain atlas (a.k.a. spatial normalization), smoothing of all voxel-values and normalization of intensities as briefly described in Section 2.2. This enables voxel-wise comparisons between images in univariate analysis (see Section 6.2.1) and the correlation (or interpretation of covariance) of all voxels within the whole-brain region in multivariate analysis.

After preprocessing, neuroimaging data is commonly reduced to a lower-dimensional subspace in the studies reviewed in this work. In most cases, this is achieved by PCA implemented as in Section 3.1, but also by the scaled subprofile model (SSM), which is a modification of PCA described in Section 3.2. Partial least squares correlation or regression (PLSC/ PLSR) is also related to PCA as it is based on the same decomposition procedure (Section 6.2.2).

The method to be used for dimensionality reduction and further analysis depends on the purpose of the study, and also on different criteria regarding stability of the dimensionality reduction. In Section 5, some criteria for the validation of PCA regarding stability and robustness are presented.

After PCA is accomplished on the neuroimaging data of AD patients and a CTR group, the resulting projections of all subjects can be used to train discrimination as described in Section 4.3. Employing MCI cases where AD is prognosed or was already confirmed, the disrimination can then be tested regarding its potential to detect AD in early stages.

A detailed outline of all methods presented in this review and a workflow for the analysis of PET and SPECT data is depicted in Figure 1.

2. Constitution of the data matrix

In all studies reviewed in this work, PET or SPECT images of asymptomatic controls and patients with AD are considered.

Both techniques generate three-dimensional images of the brain, depicting the aggregation of a radioactive tracer and therefore providing metabolic information (e.g., glucose metabolism, brain perfusion or plaque deposition) within distinct brain areas. Although PET produces images with higher resolution, SPECT is considered to be adequate to detect abnormalities of perfusion which are specific for AD (e.g., Caroli et al. (2007); Herholz et al. (2002b); Ishii et al. (1999); Matsuda (2007)). As SPECT is – in comparison to PET – more prevalent and economical, it is commonly the preferred imaging method according to Minati et al. (2009).

Overall, three tracers were used for the SPECT and PET data examined in this review:

SPECT-scans based on the tracer technetium-99m-ethyl cysteinate dimer (^{99m}Tc-ECD) show perfusion patterns of the brain. In Herholz et al. (2002b) it is observed, that superior results regarding the detection of AD and the assessment of affected brain regions can be achieved by ^{18}F-2-fluoro-2-deoxy-D-glucose (FDG) PET-imaging, which measures the changes in glucose metabolism (Ishii et al. (1999)). The tracer ^{11}Carbon-Pittsburgh compound B (^{11}C PiB) is able to quantify β-amyloid deposition in the diseased brain as pointed out by Klunk et al. (2004).

2.1 Sample selection

If a groupwise comparison of subjects with AD and CTR is intended by statistical analysis of PET or SPECT images, not all datasets are apt to be included in the sample. Especially the CTR group should be gender- and age-matched to account for age-related atrophy within the brain. The effect of age-related changes of the brain on multivariate analysis such as PCA is discussed in Zuendorf et al. (2003), where at least two principal components, i.e., two independent directions of variance, could be correlated with age.

Subjects representing the AD group should not be affected by other neuro-degenerative diseases, and are also recommended to be in a stage of mild to moderate AD. Cases of late AD, where almost the whole brain is affected, would put too much emphasis on regions still unaffected by early-onset AD.

2.2 Preprocessing of the images

In each study, the PET or SPECT images selected for statistical analysis are registered to an atlas of the brain (a.k.a. spatially normalized), smoothed and intensity normalized. An optimized preprocessing method for SPECT images is presented in Merhof et al. (2011), where a dataset containing AD and CTR subjects is preprocessed by various methods, and subsequently tested by PCA and discrimination analysis. Best results regarding robustness and classification accuracy are achieved by affine registration (Bradley et al. (2002); Herholz et al. (2002a)), smoothing of voxel intensity values based on the standard isotropic Gaussian filter with full width half maximum (FWHM) of 12mm (Herholz et al. (2002a); Ishii et al.

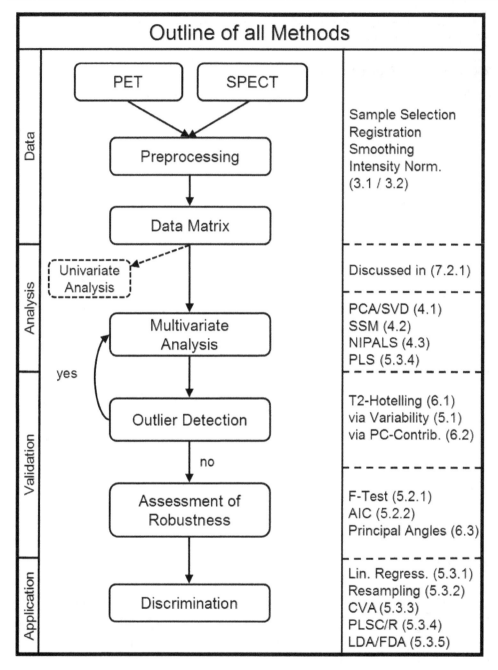

Fig. 1. Application flow and methods presented in this review

(2001); Matsuda et al. (2002)) and normalization according to the 25% brightest voxels within the whole-brain region.

To our knowledge, a detailed review of preprocessing methods and their impact on PCA applied to PET images (and subsequent analysis, with regard to discrimination of AD and CTR) has not yet been published. However, Herholz et al. (2004) present a detailed and effectual survey of the general handling of PET images in neuroscience.

After sample selection and preprocessing, the voxel-values of each scan are converted into a vector and all datasets are stored column-wise in a data matrix X as depicted in Figure 2. This enables univariate (i.e., voxel-wise) comparison, or multivariate analysis (e.g., PCA and in some cases subsequent discriminant analysis), as the observations for each voxel or brain region are now represented row-wise in X.

3. Principal component analysis

Two main implementations of PCA are considered in this review:

- The first and widely used approach is based on variance, where principal components (PCs) are determined by singular value decomposition (SVD) of the $m \times n$ data matrix X (e.g., Markiewicz et al. (2009); Merhof et al. (2011)). In this way, it is not necessary to compute the $m \times m$ covariance matrix XX^T which is time-consuming due to the very high dimensionality m of the data (in SPECT and PET images, the whole-brain region contains more than 10^5 voxels) and might even lead to a loss of precision.
- In a second implementation, PCA is modified to a scaled subprofile model (SSM) (e.g., in Habeck et al. (2008); Scarmeas et al. (2004)). SSM is also covariance-based, but also captures the regional patterns of brain function and thereby advances subsequent discriminant analysis. PCA is performed, and afterwards subject scaling factors are calculated to convey each subject's contribution to a fixed PC as described in Alexander & Moeller (1994) and Moeller et al. (1987).

Another framework is presented in Miranda et al. (2008) and Duda et al. (2001), where an approximation of PCA is achieved by minimizing the mean square error of approximation, also characterized as a total least squares regression problem (Van Huffel (1997)). However, to our knowledge this method has not been applied to SPECT or PET data of patients affected by AD and a CTR group so far and is therefore not considered further in this review.

As PCA is sensitive to outliers within the data, methods to perform a more robust PCA are also considered, e.g., in Serneels & Verdonck (2008). However, for analysis of SPECT or PET images the underlying data usually contains a manageable amount of subjects and can therefore be sorted manually or by applying tests as presented in Section 5. It is also advisable to visualize those PCs intended to remain in the subsequent analysis as explained in Section 3.4. Thereby, it can be assured that only those regions of the brain which explain the difference to CTR in mild to moderate AD are considered, and that there are no abnormally prominent regions identified by the PCA.

In this review, PCA via SVD and SSM are presented in Sections 3.1 and 3.2. During resampling, both of these methods may become unstable; therefore, an alternative implementation is indicated in Section 3.3.

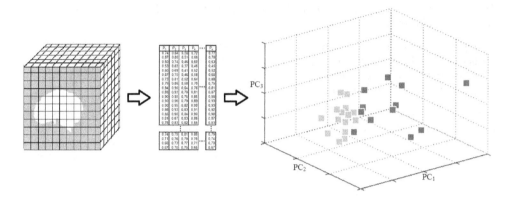

Fig. 2. Exemplary development of PCA on neuroimaging data. *Left*: Volume dataset. *Middle*: Data matrix X containing one volume dataset per column. *Right*: Projection into subspace spanned by three PCs.

A general outline of the PCA on neuroimaging data is depicted in Figure 2, where each image contained column-wise in the data matrix X is projected into a subspace spanned by the first three PCs.

3.1 PCA via singular value decomposition

Prior to multivariate analysis, the overall mean of the data matrix X is usually set to zero by subtracting the mean vector from each column. This is not compulsive but considerably simplifies further analysis (Habeck et al. (2010); Miranda et al. (2008)).

Singular value decompositon (SVD) of the data matrix is applied by $X = VSU^T$ (as in Markiewicz et al. (2009; 2011b)). As X is of size $m \times n$ with $m >> n$, it is sufficient to compute only the first n columns of V, i.e., the first n PCs. If the datasets contained in X were mean-centered beforehand, X is of rank $n - 1$ at most, so the number of PCs to be computed is furthermore reduced to $n - 1$ (this follows from the properties of the associated centering matrix, i.e., it is idempotent and therefore of rank $n - 1$).

The columns of V are sorted according to the magnitude of their associated singular values, i.e., the diagonal elements of S. PC scores for all subjects are computed by $V^T X$, i.e., each subject is projected into a PC-subspace as depicted in Figure 2. If all PCs were used, all variance of the data would be maintained, but a subset of only a few PCs is sufficient to represent more than 60% of the variance (see Section 4.2).

3.2 PCA modified to scaled subprofile model analysis

Scaled subprofile model (SSM) analysis enhances the discriminative powers of the PCA as it not only extracts the covariance structure within groups but also assesses the contribution of each subject to the covariance pattern. As explained in detail in Alexander & Moeller (1994), the data matrix X is natural log-transformed, and subsequently mean-centered over brain regions and subjects. Then PCA is performed on X as in Section 3.1 by $X = VSU^T$ and n PCs are contained in V. Furthermore, PCA via eigenvalue decomposition of the $n \times n$

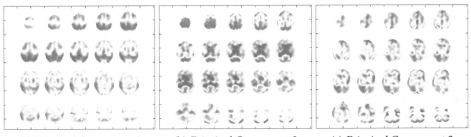

(a) Principal Component 1   (b) Principal Component 2   (c) Principal Component 3

Fig. 3. Examples for the first three principal components of a dataset containing SPECT images of 23 asymptomatic controls and 23 patients with Alzheimer's disease.

covariance matrix $X^T X$ is applied, resulting in n eigenvectors which represent sets of subject scaling factors (SSFs). The associated eigenvalues to the PCs, SSFs respectively, of both decompositions are equal. Whereas the PCs describe the main directions of variance in the data, the SSFs describe the degree of subjects' expression of the fixed PC (Habeck et al. (2008)). The expression of the PC-scores $V^T X$ for each subject is quantified by the associated SSFs in accordance with the procedure described in Alexander & Moeller (1994) and Habeck et al. (2008). As above in Section 3.1, only a few PCs and associated SSFs are sufficient to reflect pathological differences within the data.

3.3 PCA via non-linear iterative partial least squares

During bootstrap resampling (e.g., to assess robustness of the PCA as described in Section 5.3), individual subjects may be repeatedly present within the resampled data matrix, thereby rendering the SVD unstable.

In this case, Markiewicz et al. (2009) propose the application of the non-linear iterative partial least squares (NIPALS) algorithm. The (resampled) data matrix X is decomposed by $X = v_1 t_1^T + R$, where v_1 denotes an estimate of the first PC of X, t_1 represents the appendent PC scores of each subject and R is the remaining residual. As an estimate for v_1, Wold et al. (1987) propose the (normalized) column of X with the largest variance, but the employment of other start vectors is possible as well (Miyashita et al. (1990)). The NIPALS algorithm is iterated with R acting as new start matrix until all PCs required for further analysis are computed. The NIPALS method is related to canonical correlation analysis (Höskuldsson (1988)), and thereby also to canonical variate analysis as presented in Section 4.3.3.

3.4 Visualization of PCs

Axial slices of PCs can be visualized as illustrated in Figure 3, where ^{99m}Tc-ECD SPECT images of 23 subjects with Alzheimer's disease and 23 asymptomatic controls were decomposed by PCA via SVD. As PCA seeks directions for representation (rather than discrimination), the displayed patterns are not to be mistaken with discriminant images.

The voxel-values of each PC are converted back into a three-dimensional image (reverse to the procedure in Section 2.1), and every third slice of the PC-image between slice 15 and 72 is depicted. The intensities of the voxel-values are mapped to a colormap ranging from blue negative values to red positive values. Neutral voxel-values ($= 0$, as the data was mean-centered) correspond to white.

The main variance observed in the temporal lobes is captured in the first PC, whereas the second PC expresses changes in the area of the ventricles, which could be attributed to the expansion of ventricles in AD patients. A first intuitive conclusion might be to maintain only the first two PCs for further analysis, as those describe the regions usually affected by AD within the brain most distinctly. However, there are more reliable methods to decide which PCs to keep (see Section 4.2).

4. Applications

In the statistical evaluation of neuroimages, the main purpose of PCA is primarily an efficient reduction of the very high dimensionality and the removal of noise and redundant information within the data. The PCs produced during PCA represent the axes of the new subspace, into which the original datasets containing the voxel-values are transformed. The decision which PCs are suited to represent the data sufficiently has a great impact on all further analysis. Therefore, the contribution of each PC should be thoroughly evaluated. Different criteria for choosing a well-fitting subset of PCs are presented in Section 4.2. Also, the measurement of the amount of variability maintained within each PC is closely connected with the question of its significance (see also Section 4.1).

If the dataset at hand contains two (or more) groups of subjects, the PCs established to be relevant for further analysis are found to notably describe those regions within the brain, which differ significantly across groups. PCA can therefore be useful to train a discrimination or to provide the basis for subsequent discriminant analysis as presented in Section 4.3.

4.1 Explanation of the variability

Under the condition that the variables (voxels) of all subjects are on the same scale (this has to be ensured during preprocessing of the images), the variance of the ith PC equals its associated eigenvalue e_i (Massy (1965)). The percentage of the accumulated variance represented by any number n of all N PCs is then calculated by

$$var(n) = \frac{\sum_{i=1}^{n} e_i}{\sum_{i=1}^{N} e_i}. \tag{1}$$

In several studies it is observed that the first few PCs generally account for more than 60% of the variability (e.g., in Habeck et al. (2008); Markiewicz et al. (2009)). The percentage of the cumulative variance explained is used by Fripp et al. (2008a) to compare different methods for preprocessing of the data, e.g., spatial registration to different brain atlases.

4.2 Dimensionality reduction

In neuroimaging, the number of variables m (i.e., voxels of the whole-brain region) greatly outnumbers the number of observations n (i.e., subjects included in the study). For this reason, a dimensionality reduction of the data before further analysis, such as discrimination or correlation (as in Pagani et al. (2009)), is commonly applied. PCA is well suited for this purpose, as it reduces the variable space to a few dimensions only. It also helps to focus on the main directions of variance within the data (i.e., the first few PCs) and treats unused PCs corresponding to lower eigenvalues as noise in the data.

In each of the reviewed studies, only the first few principal components (PCs) are used to represent the main variance of the data. In some cases, this is justified by execution of the partial F-test as presented in Section 4.2.1(Markiewicz et al. (2009)), by calculation of the cumulative variance explained by the PCs (e.g., Fripp et al. (2008a); Zuendorf et al. (2003), see also Section 4.1) or by application of Akaike's information criterion (Habeck et al. (2008); Scarmeas et al. (2004), see also Section 4.2.2).

4.2.1 Partial F-test

The partial F-test measures which PCs add significant variance to the data (Markiewicz et al. (2009)). In the beginning, the mean-centered data matrix $X = X_{start}$ is entered into a regression model, and its residual sum of squares $rss(1)$ is computed. In a first iteration, the first PC v_1 is added to the model and prediction values for the original data matrix are calculated by $v_1 v_1^T X_{start}$. Then the residual sum of squares of the deviation matrix $D = X_{start} - v_1 v_1^T X_{start}$ is calculated. In each of the following $N - 1$ iterations, D and the next PC are entered into the model.

F-values and p-values for each iteration are calculated by

$$F_n = \frac{(rss(n) - rss(n+1))(N - n)}{rss(n+1)} \tag{2}$$

and

$$p_n = 1 - fcdf(F(n)), \tag{3}$$

where $fcdf$ denotes the F cumulative distribution function with numerator and denominator degrees of freedom 1 and $N - n - 1$.

As the limiting factor for number of PCs, Markiewicz et al. (2009) propose p to be lower than 0.05, which is a standard level of significance.

4.2.2 Akaike's information criterion

Similar to partial F-test, Akaike's information criterion (AIC) determines the subset of PCs which represents the best fitting model (Akaike (1974)).

AIC-values are calculated by

$$A = -2log(L) + 2K, \tag{4}$$

where L denotes the maximum value of the log-likelihood function of the model and K the number of estimable parameters (Burnham & Anderson (2002)). The model which scores the smallest AIC-value A is considered to be the best fitting one. As AIC may be biased if the ratio of sample size and number of parameters is too small (e.g., $\frac{n}{K} < 40$), Sugiura (1978) proposes a correction factor (AIC$_c$):

$$A_c = A + \frac{2K(K+1)}{n - K - 1}. \tag{5}$$

Burnham & Anderson (2002) recommend the usage of AIC$_c$ in any case, as AIC and AIC$_c$ are similar for a sufficiently large ratio $\frac{n}{K}$.

In Habeck et al. (2008), AIC-values A are calculated only for models generated by the first six PCs (explaining more than 75% of all variance), and the best-fitting model with the lowest

AIC-value is chosen for subsequent analysis. However, it should be noted that the AIC does not recognize if none of the models is suited to represent the population, i.e., the PCs entered into the AIC need to be chosen carefully.

4.3 Discrimination methods

With regards to the early detection of AD, the discriminative power of PCA can be very valuable. Discrimination should be trained on subjects with mild to moderate AD and asymptomatic CTR, and afterwards be tested on MCI cases, thereby assessing the capability to detect early AD cases among the data collected for the study.

Due to the orthogonality of all eigenvectors, each PC is uncorrelated with all preceding PCs and therefore captures an independent feature of the dataset. As the main variance resides in the first PCs, they depict prominent features of the data (provided that there are no outliers). Hence, the PCs can be employed for the differentiation of groups within the dataset. Those PCs which best discriminate the subjects can either be determined in a linear regression model as presented in Section 4.3.1 (Habeck et al. (2008); Scarmeas et al. (2004)) or as in Section 4.3.2 by a leave-one-out approach (Fripp et al. (2008a)). If necessary, discrimination can be refined further, e.g., by Canonical variate analysis (Section 4.3.3) or Fisher's discriminant analysis (Section 4.3.5).

4.3.1 Linear regression

Linear regression is a subtype of general regression analysis and is widely used for the identification of those independent variables, which relate strongly to the dependent variable (e.g., group membership). After the successful completion of the regression, it can furthermore be applied to predict the group membership of a newly added value.

The achieved PC-scores $\tilde{X}$ of each subject are entered as independent variables into the linear regression model $y = \tilde{X}b + \epsilon$. The vector y of the subjects' group memberships, in this case AD and CTR, contains the dependent variables.

It is common to use only a subset of all PCs (determined by significance tests or the amount of variance they represent), but it should be noted that even a PC which captures little variance might still be related to a dependent variable (Jolliffe (1982)).

The regression results in a linear combination of those PCs which achieve the best differentiation of the two classes (e.g., Habeck et al. (2008); Scarmeas et al. (2004)).

If the dependent variables include more information than group membership (e.g., age or existence of genetic risk factors), partial least squares (PLS) regression can be applied (see also Section 4.3.4). This method generalizes PCA and multiple linear regression.

4.3.2 Leave-one-out resampling

In leave-one-out resampling, one subject is drawn from the underlying data sample per iteration and subsequent analysis is applied. This measures the individual contribution of each subject and can therefore be applied to sort out abnormal interference of particular subjects where necessary.

In Fripp et al. (2008a), $n - 1$ out of n images are decomposed by PCA in each iteration. Then, PC-scores of the subjects contained in the sample are plotted pairwise against each other.

Those PCs which generally provide the best cluster and separation of the groups within iterations are considered for further analysis.

4.3.3 Outline of canonical variate analysis

Canonical variate analysis (CVA) is another regression model considered to enhance the discriminative strength of PCA in neuroimaging. Similarly to linear regression, it identifies the best separation of groups depending on PC-scores. The first canonical variable (CV) is the best of all possible linear combinations of PC-scores for differentiation of the groups and – analogous to PCs – the following CVs are computed under the condition to be orthogonal to all precedent CVs.

PCA is applied for dimensionality reduction and removal of noise (i.e., discarded PCs). The within- and between-group sum-of-squares and crossproduct matrices W and B are computed for the PC-scores of all subjects. Then the CVs, i.e., the eigenvectors of $W^{-1}B$, are linear combinations of PC-scores and are sorted by their discriminative power (Borroni et al. (2006); Kerrouche et al. (2006)). CV-scores of all subjects are calculated analogous to PC-scores.

4.3.4 Outline of partial least squares correlation and regression

As in PCA, the main element of partial least squares (PLS) methods is the SVD, which is applied to the correlation matrix YX^T (rather than the data matrix X containing the mean-centered data, as in PCA). The independent variables are the mean-centered and normalized voxel-values of all brain images stored in X, and the n vectors of dependent variables for all subjects (also mean-centered and normalized) form the $k \times n$ matrix Y. SVD of YX^T produces VSU^T, where S is a diagonal matrix containing singular values and U and V column-wise contain the left respectively right singular vectors. Analogous to PCA, it is sufficient to compute only the first few columns of V. Then, the high-dimensional data of X is reduced by $T = X^TU$ (a.k.a. brain scores), and Y is reduced to Y^TV (a.k.a. behavior scores).

It depends on the intention of the study, in which way these results are further analysed and applied. Krishnan et al. (2011) give an elaborate survey of the main PLS methods used in neuroimaging as well as of practical implementations. Generally, they present two basic approaches, i.e., PLS regression and PLS correlation. PLS regression is a generalization of multiple linear regression and PCA (Abdi (2010)), and is used to predict behavior on the basis of neuroimages, in this case PET or SPECT data. PLS correlation focuses on the analysis of the relation between two groups within the dataset and can be subdivided into more specific applications according to the design of the research.

4.3.5 Outline of linear and Fisher's discriminant analysis

Similar to CVA, linear discriminant analysis (LDA) seeks discriminative directions of the data rather than representative directions (as does PCA). It can be applied both to the original mean-centered voxel-values contained in the data matrix X or in a second step after performance of PCA to the PC-scores of all subjects. The latter approach is preferable when dealing with high-dimensional data, as either the inverse of an $m \times m$ scatter matrix has to be computed or a generalized eigenvalue decomposition of $m \times m$ matrices is required.

Fisher's discriminant analysis (FDA) is a special application of LDA, without the constraints of normal distributed groups and equal group covariance. It has lately been applied several times to diffentiate between subjects with AD and normal controls, e.g., in Markiewicz et al. (2009; 2011a); Merhof et al. (2009; 2011).

The purpose of FDA is to maximize the ratio of the between- and the within-group scatter S_B and S_W, thereby projecting the data into a one-dimensional subspace. This is achieved by the projection vector w, i.e., the solution of the generalized eigenvalue problem $S_W^{-1} S_B w = \lambda w$ (Duda et al. (2001)). Subsequent classification can be computed with very limited effort by a threshold or nearest-neighbor approach.

5. Derivation of robustness of the PCA

So far, PCA and its applications in neuroimaging were introduced, but not yet validated and discussed. It is very important to assess the robustness of the PCA (and, where necessary, subsequent procedures) before interpretation of the results, as instability and overtraining may occur for various reasons. PCA is sensitive to conspicuous cases and it is therefore recommended to inspect the resulting PCs before further analysis. In order to ensure that no pathologically abnormal cases (outliers) remain in the training set, the T^2-Hotelling test is executed, e.g., by Pagani et al. (2009); Zuendorf et al. (2003) (see Section 5.1). Kerrouche et al. (2006) also propose further measurement of the individual contribution of one observation to each PC (see Section 5.2), to assess if the removal of one observation changes the outcome of PCA significantly. Habeck et al. (2010) also observe that if the first PC contains more than 90% of the variance to the data, it is very probable that the dataset X includes one or more outliers (see Section 4.1).

By bootstrap resampling of the dataset and subsequent PCA the instability caused by removal of a subset of subjects is measured (Markiewicz et al. (2009; 2011a); Merhof et al. (2011)) via principal angles between PC-subspaces.

5.1 Hotelling's T-square test

Hotelling's T-square test is an adaption of the Student's T-test to the multivariate case (Hotelling (1931)). As the F-distribution is more prevalent, the T^2-distribution is usually transformed to

$$T^2 \sim \frac{p(n-1)}{n-p} F_{p,n-p}, \tag{6}$$

where n denotes the number of subjects and p the number of PCs retained in the model. Let y_i denote the column vector of PC-scores of the ith subject, then its T^2-value is obtained by $T^2 = y_i^T y_i$. Zuendorf et al. (2003) propose a threshold of $p < 0.01$, and further assess the validity of the T^2-test by adding an abnormal case to a set of normal controls in 15 iterations. However, the T^2-test can also be applied to a dataset containing two or more groups (Kerrouche et al. (2006); Pagani et al. (2009)).

5.2 Contribution of subjects to PCs

The amount of the contribution $c_{i,j}$ of the ith subject to the jth PC is measured by

$$c_{i,j} = \frac{1}{n-1} \cdot \frac{y_i(j)}{e_j}, \tag{7}$$

where n denotes the number of all subjects, y_i the column vector of PC-scores of the ith subject and e_j the eigenvalue corresponding to the jth PC. An abnormally large value of $c_{i,j}$ indicates that the removal of the ith subject might significantly change the results of PCA (Kerrouche et al. (2006)).

5.3 Principal angles of PC-subspaces

In order to compare sets of PCs during resampling iterations, the use of principal angles between PC-subspaces of a fixed dimension is proposed by Markiewicz et al. (2009). If the largest principal angle between an original and resampled subspace is very small, the PCA can be considered to be sufficiently independent of the underlying training set. Otherwise, abnormal large principal angles can indicate that too many PCs (i.e., too much noise) are included in the analysis, or that the sample was not selected carefully enough with respect to outliers.

In bootstrap resampling, n subjects are drawn with replacement from the original training set (Efron & Tibshirani (1993)). For better replication of the original set, AD and CTR cases are stratified in the bootstrap sample (Markiewicz et al. (2009)).

For every sample, PCA is performed and the subspace spanned by the first i PCs is compared to the i-dimensional PC-subspace of the original set. This is achieved by calculating the largest principal angle between the two subspaces (Golub & van Van Loan (1996); Knyazev et al. (2002)). For any number of PCs, the mean angle across all iterations is computed. Increased angles indicate the destabilization of the PCA.

The same method can be applied with leave-one-out resampling, i.e., one subject of each group is dropped in every iteration (Markiewicz et al. (2009)).

The computation of prinipal angles should be treated with caution, as round-off errors might cause inaccurate estimates for small angles. A solution to this problem is proposed by Knyazev et al. (2002), where a combined sine and cosine based approach is presented and generalized.

6. Discussion

For analysis of SPECT and PET data, PCA is widely applied and commonly reported to deliver stable and efficient results when used correctly. However, some limitations of PCA outlined in Section 6.1 remain, which might interfere strongly with the outcome of the statistical analyses. In some cases it might even be advisable to apply alternative methods to obtain more reliable results. In Section 6.2, examples are given where the performance of PCA on neuroimaging data was investigated and compared to other approaches.

6.1 Limitations of the PCA in neuroimaging

As PCA is based solely on the decomposition of the covariance matrix, the underlying data must be dealt with carefully. The preprocessing of the images has a crucial impact on the outcome of the analysis as pointed out by Fripp et al. (2008a), where PCA on ^{11}C PiB PET data proved to be sensitive to inaccuracies originating from non-rigid registration and intensity normalization. On ^{99m}Tc-ECD SPECT, the classification accuracy of AD and CTR subjects via PCA and subsequent FDA depends significantly on the data preprocessing method (Merhof et al. (2011)). Classification accuracy relies also on scanner type and reconstruction method of FDG-PET, if the data is aquired from a more heterogeneous dataset, e.g., from the ADNI database as in Markiewicz et al. (2011b).

Not only the preprocessing of the training set but also its composition is essential. This includes the stratification of groups, the sample sizes and the absence of outliers as described in Sections 6.1.1 and 6.1.2. Moreover, the number of PCs retained in the analysis is important and depends on the purpose of the study (see Section 6.1.3).

6.1.1 Sample size

The selection of subjects suited for training is constrained by many premises, such as age- and gender-matched CTR cases, stage of AD, the absence of other neurodegenerative disorders and the quality of the scan. It is also preferable for all images to be aquired by the same scanner, as this improves comparability of the data. Therefore, most studies only include less than 30 subjects of each group, except for Markiewicz et al. (2011b) where previous results on a smaller and more homogeneously selected training set were validated on more than 160 AD and CTR cases obtained from the ADNI database. ADNI provides generally accessible data of patients diagnosed with AD or MCI and of normal controls collected from various clinical sites (Mueller et al. (2005)).

An under-sized training set might cause the extraction of instable features (Markiewicz et al. (2009)) resulting in overly optimistic results of subsequent analysis (Markiewicz et al. (2011a)). This might be remedied by bootstrap resampling of the training set but must rely on the assumption that the sample is representative of the population (Markiewicz et al. (2009)).

6.1.2 Sensitivity to outliers

As the covariance matrix is calculated empirically, the estimates of eigenvectors (PCs) are heavily influenced by outliers, i.e., pathologically abnormal cases within the training dataset. The variance caused by only one outlier may be captured within the first PC, which will thereby not regard the variance within regular cases and dramatically change further results. Approaches which substitute the original covariance matrix by a more robust estimate (e.g., in Debruyne & Hubert (2006)) exist, but these methods are not practical for datasets of high dimensionality. For this reason, it is highly recommended to determine outliers by additional testing when applying PCA to neuroimaging data.

6.1.3 Number of PCs

Although several approaches are presented in Section 4.2, the determination of how many and which PCs are best suited to represent the original images remains subject to interpretation. So far, much relies on the purpose of PCA and the further analysis of the data. An elaborate overview over criteria for estimating the number of significant PCs and their application is presented in Peres-Neto et al. (2005), and some of these apply to covariance matrices as well as correlation matrices. The application of such methods can be ambivalent, as reviewed by Franklin et al. (1995). In most studies, PCs are chosen according to their potential to explain data and their impact on robustness. It is therefore advisable to determine the number of retained PCs not only based on one criterion but also on the best possible trade-off between the resulting accuracy and robustness.

6.2 Comparison to similar methods

In Section 4, different extensions to PCA such as linear regression or canonical variate analysis are presented, and also an outline of methods with similar properties or intentions as PCA is given. The decision which of these methods is best to employ always depends on the underlying research question, on the data available and the selected sample.

Sections 6.2.1 to 6.2.3 provide a review of the most important methods frequently applied to neuroimaging data and compare them to results obtained from PCA.

6.2.1 Univariate analysis of neuroimaging data

Univariate analysis measures the voxel-by-voxel correlation between groups (e.g., by a voxel-wise T-test in Habeck et al. (2008)) and thereby merely focuses on the identification of significant shifts between voxel-values. In neuroimaging, univariate methods are commonly used during image preprocessing, e.g., in Dukart et al. (2010) or Scarmeas et al. (2004) for intensity normalization. Voxel-wise analysis can also be used for differentiation between groups, but it is unanimously reported that multivariate approaches outperfom univariate analysis in this matter, especially in the detection of early-onset cases of dementia (Habeck et al. (2008); Scarmeas et al. (2004)). Another drawback of voxel-wise analysis is the sensitivity regarding the preprocessing of the data, and even under the assumption that an optimized normalization factor was applied, the interpretion of the results must be addressed in a multivariate fashion (McCrory & Ford (1991)). In another approach, Higdon et al. (2004) tried to apply a between-group T-test for dimensionality reduction, but this proved to be ineffective and even deteriorated accuracy results.

On the other hand, multivariate analysis is found to be more robust as it considers the entire covariance structure of the data (accounting for relations among regions) and withstands the deviation of individual voxel-values (Borroni et al. (2006); Habeck et al. (2008)). It thereby detects correlated alterations in a diseased brain, whereas univariate analysis might not be able to recognize these differences.

6.2.2 Partial least squares

When examining very high-dimensional data, and especially for the discrimination into groups within the dataset, PLS has been reported to perform better than PCA regarding the classification accuracy (Higdon et al. (2004); Kemsley (1996)). This is rather self-evident, as PCA does not take into account further behavioral data (e.g., neuropsychological data such as Mini-Mental State Examination (MMSE) scores, age, years of education). If PLS is applied for dimensionality reduction, Kemsley (1996) reports that fewer PLS dimensions than PCs were required for a successful subsequent differentiation of the groups. This implies that PLS will capture the most discriminative attributes of the subjects within the first dimensions, rather than the representative directions generated by PCA.

Nevertheless, there are certain drawbacks in the application of PLS methods. PLS tends to overfit the data, so the determination of the number of PLS dimensions kept in the analysis is of decisive importance (Abdi (2010)). In addition, PLS may detect differences which are not characteristic of the examined groups but were produced randomly by noise within the underlying dataset (Kemsley (1996)). Furthermore, PLS only works under the assumption that behavior relates linearly to neuroimaging data (McIntosh & Lobaugh (2004)).

Overall, if allowances are made for these effects and significant behavioral data is available, PLS can still be a favourable alternative to PCA.

6.2.3 Linear discriminant analysis

As explained above, performance of PCA (or any other dimensionality reduction method) prior to LDA is preferable in neuroimaging due to the high dimensionality of the data and the resulting expensive computation. To our knowlegde, LDA as described above in Section 4.3.5 has not yet been applied to discriminate AD from CTR using voxel-values of PET or SPECT images of the whole-brain region, although McEvoy et al. (2009) utilize a stepwise approach of LDA to identify brain regions significant for differentiation.

In other areas also dealing with high-dimensional data, such as object recognition in images, LDA is usually considered to perform superior to PCA. But this is not necessarily the case for small training sets, as pointed out by Martínez & Kak (2001). In the same study they also observe PCA to be less biased than LDA, i.e., less constrained to the training set.

The overall good results regarding accuracy and robustness of the PCA-LDA or PCA-FDA approach (e.g., as presented in Markiewicz et al. (2009)) also indicate, that a preceding PCA does not impair the discriminant analysis.

7. Conclusion

PCA applied to SPECT or PET data is well suited to reduce the high dimensionality of the original dataset containing voxel-values of the whole-brain region. It achieves best results when data is transformed into a subspace spanned by a well-chosen subset of PCs that represents the variability within all datasets and at the same time reduces noise and redundant information. PCA can also be used successfully to train discrimination between AD and a set of asymptomatic CTRs with the intention to enable an early detection of AD, or to provide a stable and effective basis for the subsequent application of discriminant analysis.

8. References

Abdi, H. (2010). Partial least squares regression and projection on latent structure regression (PLS Regression), *Wiley Interdisciplinary Reviews: Computational Statistics* 2(1): 97–106.

Akaike, H. (1974). A new look at the statistical model identification, *IEEE Transactions on Automatic Control* 19(6): 716–723.

Alexander, G. E. & Moeller, J. R. (1994). Application of the scaled subprofile model to functional imaging in neuropsychiatric disorders: A principal component approach to modeling brain function in disease, *Human Brain Mapping* 2(1-2): 79–94.

Borroni, B., Anchisi, D., Paghera, B., Vicini, B., Kerrouche, N., Garibotto, V., Terzi, A., Vignolo, L., Luca, M. D., Giubbini, R., Padovani, A. & Perani, D. (2006). Combined 99mTc-ECD SPECT and neuropsychological studies in MCI for the assessment of conversion to AD, *Neurobiology of Aging* 27(1): 24 – 31.

Braak, H. & Braak, E. (1991). Neuropathological stageing of Alzheimer-related changes, *Acta Neuropathologica* 82: 239–259.

Bradley, K. M., O'Sullivan, V. T., Soper, N. D. W., Nagy, Z., King, E. M., Smith, A. D. & Shepstone, B. J. (2002). Cerebral perfusion SPET correlated with Braak pathological stage in Alzheimer's disease, *Brain* 125(8): 1772–1781.

Burnham, K. P. & Anderson, D. R. (2002 2nd edn). *Model Selection and Multi-Model Inference. A Practical Information-Theoretic Approach*, New York Springer-Verlag.

Caroli, A., Testa, C., Geroldi, C., Nobili, F., Barnden, L., Guerra, U., Bonetti, M. & Frisoni, G. (2007). Cerebral perfusion correlates of conversion to Alzheimer's disease in amnestic mild cognitive impairment, *Journal of Neurology* 254: 1698–1707.

Carr, D., Goate, A., Phil, D. & Morris, J. (1997). Current Concepts in the Pathogenesis of Alzheimer's Disease, *The American Journal of Medicine* 103(3, Supplement 1): 3S – 10S.

Debruyne, M. & Hubert, M. (2006). The Influence Function of Stahel-Donoho Type Methods for Robust Covariance Estimation and PCA.

Duda, R. O., Hart, P. E. & Stork, D. G. (2001). *Pattern Classification*, Wiley-Interscience.

Dukart, J., Mueller, K., Horstmann, A., Vogt, B., Frisch, S., Barthel, H., Becker, G., Möller, H. E., Villringer, A., Sabri, O. & Schroeter, M. L. (2010). Differential effects of global and cerebellar normalization on detection and differentiation of dementia in FDG-PET studies, *NeuroImage* 49(2): 1490 – 1495.

Efron, B. & Tibshirani, R. J. (1993). *An Introduction to the Bootstrap*, Chapman & Hall/CRC.

Franklin, S. B., Gibson, D. J., Robertson, P. A., Pohlmann, J. T. & Fralish, J. S. (1995). Parallel analysis: a method for determining significant principal components, *Journal of Vegetation Science* 6(1): 99–106.

Fripp, J., Bourgeat, P., Acosta, O., Raniga, P., Modat, M., Pike, K. E., Jones, G., O'Keefe, G., Masters, C. L., Ames, D., Ellis, K. A., Maruff, P., Currie, J., Villemagne, V. L., Rowe, C. C., Salvado, O. & Ourselin, S. (2008a). Appearance modeling of 11C PiB PET images: Characterizing amyloid deposition in Alzheimer's disease, mild cognitive impairment and healthy aging, *NeuroImage* 43(3): 430 – 439.

Fripp, J., Bourgeat, P., Raniga, P., Acosta, O., Villemagne, V. L., Jones, G., O'Keefe, G., Rowe, C. C., Ourselin, S. & Salvado, O. (2008b). MR-less high dimensional spatial normalization of 11C PiB PET images on a population of elderly, mild cognitive impaired and Alzheimer disease patients., *Medical Image Computing and Computer-Assisted Intervention* 11(Pt 1): 442–449.

Golub, G. H. & van Van Loan, C. F. (1996). *Matrix Computations (Johns Hopkins Studies in Mathematical Sciences)*, 3rd edn, The Johns Hopkins University Press.

Habeck, C., Foster, N. L., Perneczky, R., Kurz, A., Alexopoulos, P., Koeppe, R. A., Drzezga, A. & Stern, Y. (2008). Multivariate and univariate neuroimaging biomarkers of Alzheimer's disease, *NeuroImage* 40(4): 1503–1515.

Habeck, C., Stern, Y. & (2010). Multivariate Data Analysis for Neuroimaging Data: Overview and Application to Alzheimer's Disease, *Cell Biochemistry and Biophysics* 58: 53–67. 10.1007/s12013-010-9093-0.

Herholz, K., Herscovitch, P. & Heiss, W. (2004). *NeuroPET: PET in Neuroscience and Clinical Neurology*, Springer, Berlin.

Herholz, K., Salmon, E., Perani, D., Baron, J., Holthoff, V., Frölich, L., Schönknecht, P., Ito, K., Mielke, R., Kalbe, E., Zündorf, G., Delbeuck, X., Pelati, O., Anchisi, D., Fazio, F., Kerrouche, N., Desgranges, B., Eustache, F., Beuthien-Baumann, B., Menzel, C., Schröder, J., Kato, T., Arahata, Y., Henze, M. & Heiss, W. (2002a). Discrimination between Alzheimer dementia and controls by automated analysis of multicenter FDG PET, *Neuroimage* 17(1): 302–316.

Herholz, K., Schopphoff, H., Schmidt, M., Mielke, R., Eschner, W., Scheidhauer, K., Schicha, H., Heiss, W. & Ebmeier, K. (2002b). Direct Comparison of Spatially Normalized PET and SPECT Scans in Alzheimer's Disease, *Journal of Nuclear Medicine* 43(1): 21–26.

Higdon, R., Foster, N. L., Koeppe, R. A., DeCarli, C. S., Jagust, W. J., Clark, C. M., Barbas, N. R., Arnold, S. E., Turner, R. S., Heidebrink, J. L. & Minoshima, S. (2004). A comparison of classification methods for differentiating fronto-temporal dementia from Alzheimer's disease using FDG-PET imaging, *Statistics in Medicine* 23(2): 315–326.

Höskuldsson, A. (1988). PLS regression methods, *Journal of Chemometrics* 2(3): 211–228.

Hotelling, H. (1931). The generalization of student's ratio, *The Annals of Mathematical Statistics* 2(3): pp. 360–378.

Ishii, K., Sasaki, M., Sakamoto, S., Yamaji, S., Kitagaki, H. & Mori, E. (1999). Tc-99m Ethyl Cysteinate Dimer SPECT and 2-[F-18]fluoro-2-deoxy-D-glucose PET in Alzheimer's Disease: Comparison of Perfusion and Metabolic Patterns, *Clinical Nuclear Medicine* 24(8): 572–575.

Ishii, K., Willoch, F., Minoshima, S., Drzezga, A., Ficaro, E. P., Cross, D. J., Kuhl, D. E. & Schwaiger, M. (2001). Statistical Brain Mapping of 18F-FDG PET in Alzheimer's Disease: Validation of Anatomic Standardization for Atrophied Brains, *Journal of Nuclear Medicine* 42(4): 548–557.

Jolliffe, I. T. (1982). A Note on the Use of Principal Components in Regression, *Journal of the Royal Statistical Society. Series C (Applied Statistics)* 31(3): pp. 300–303.

Kemsley, E. (1996). Discriminant analysis of high-dimensional data: a comparison of principal components analysis and partial least squares data reduction methods, *Chemometrics and Intelligent Laboratory Systems* 33(1): 47 – 61.

Kerrouche, N., Herholz, K., Mielke, R., Holthoff, V. & Baron, J.-C. (2006). 18FDG PET in vascular dementia: differentiation from Alzheimer's disease using voxel-based multivariate analysis, *J Cereb Blood Flow Metab* 26: 1213–1221.

Klunk, W. E., Engler, H., Nordberg, A., Wang, Y., Blomqvist, G., Holt, D. P., Bergström, M., Savitcheva, I., Huang, G.-F., Estrada, S., Ausén, B., Debnath, M. L., Barletta, J., Price, J. C., Sandell, J., Lopresti, B. J., Wall, A., Koivisto, P., Antoni, G., Mathis, C. A. & Långström, B. (2004). Imaging brain amyloid in Alzheimer's disease with Pittsburgh Compound-B, *Annals of Neurology* 55(3): 306–319.

Knyazev, A. V., Merico & Argentati, E. (2002). Principal angles between subspaces in an a-based scalar product: Algorithms and perturbation estimates, *SIAM J. Sci. Comput* 23: 2009–2041.

Krishnan, A., Williams, L. J., McIntosh, A. R. & Abdi, H. (2011). Partial Least Squares (PLS) methods for neuroimaging: A tutorial and review, *NeuroImage* 56(2): 455 – 475.

Markiewicz, P. J., Matthews, J. C., Declerck, J. & Herholz, K. (2009). Robustness of multivariate image analysis assessed by resampling techniques and applied to FDG-PET scans of patients with Alzheimer's disease, *NeuroImage* 46(2): 472 – 485.

Markiewicz, P., Matthews, J., Declerck, J. & Herholz, K. (2011a). Robustness of correlations between PCA of FDG-PET scans and biological variables in healthy and demented subjects, *NeuroImage* 56(2): 782 – 787. Multivariate Decoding and Brain Reading.

Markiewicz, P., Matthews, J., Declerck, J. & Herholz, K. (2011b). Verification of predicted robustness and accuracy of multivariate analysis, *NeuroImage* 56(3): 1382 – 1385.

Martínez, A. M. & Kak, A. C. (2001). PCA versus LDA, *IEEE Trans. Pattern Anal. Mach. Intell.* 23: 228–233.

Massy, W. F. (1965). Principal Components Regression in Exploratory Statistical Research, *Journal of the American Statistical Association* 60(309): pp. 234–256.

Matsuda, H. (2007). Role of Neuroimaging in Alzheimer's Disease, with Emphasis on Brain Perfusion SPECT, *Journal of Nuclear Medicine* 48(8): 1289–1300.

Matsuda, H., Kitayama, N., Ohnishi, T., Asada, T., Nakano, S., Sakamoto, S., Imabayashi, E. & Katoh, A. (2002). Longitudinal Evaluation of Both Morphologic and Functional Changes in the Same Individuals with Alzheimer's Disease, *Journal of Nuclear Medicine* 43(3): 304–311.

McCrory, S. J. & Ford, I. (1991). Multivariate analysis of spect images with illustrations in Alzheimer's disease, *Statistics in Medicine* 10(11): 1711–1718.

McEvoy, L. K., Fennema-Notestine, C., Roddey, J. C., Hagler, D. J., Holland, D., Karow, D. S., Pung, C. J., Brewer, J. B. & Dale, A. M. (2009). Alzheimer Disease: Quantitative Structural Neuroimaging for Detection and Prediction of Clinical and Structural Changes in Mild Cognitive Impairment1, *Radiology* 251(1): 195–205.

McIntosh, A. R. & Lobaugh, N. J. (2004). Partial least squares analysis of neuroimaging data: applications and advances, *NeuroImage* 23, Supplement 1(0): S250 – S263.

Merhof, D., Markiewicz, P. J., Declerck, J., Platsch, G., Matthews, J. C. & Herholz, K. (2009). Classification Accuracy of Multivariate Analysis Applied to 99mTc-ECD SPECT Data in Alzheimer's Disease Patients and Asymptomatic Controls, *Nuclear Science Symposium Conference Record (NSS/MIC)* pp. 3721–3725.

Merhof, D., Markiewicz, P. J., Platsch, G., Declerck, J., Weih, M., Kornhuber, J., Kuwert, T., Matthews, J. C. & Herholz, K. (2011). Optimized data preprocessing for multivariate analysis applied to 99mTc-ECD SPECT data sets of Alzheimer's patients and asymptomatic controls, *J Cereb Blood Flow Metab* 31(1): 371–383.

Minati, L., Edginton, T., Grazia Bruzzone, M. & Giaccone, G. (2009). Reviews: Current Concepts in Alzheimer's Disease: A Multidisciplinary Review, *American Journal of Alzheimer's Disease and Other Dementias* 24(2): 95–121.

Miranda, A. A., Borgne, Y.-A. & Bontempi, G. (2008). New Routes from Minimal Approximation Error to Principal Components, *Neural Process. Lett.* 27: 197–207.

Miyashita, Y., Itozawa, T., Katsumi, H. & Sasaki, S.-I. (1990). Comments on the NIPALS algorithm, *Journal of Chemometrics* 4(1): 97–100.

Moeller, J. R., Strother, S. C., Sidtis, J. J. & Rottenberg, D. A. (1987). Scaled Subprofile Model: A Statistical Approach to the Analysis of Functional Patterns in Positron Emission Tomographic Data, *Jornal of Cerebral Blood Flow & Metabolism* 7(5): 649–658.

Mueller, S. G., Weiner, M. W., Thal, L. J., Petersen, R. C., Jack, C., Jagust, W., Trojanowski, J. Q., Toga, A. W. & Beckett, L. (2005). The Alzheimer's Disease Neuroimaging Initiative, *Neuroimaging Clinics of North America* 15(4): 869 – 877.

Pagani, M., Salmaso, D., Rodriguez, G., Nardo, D. & Nobili, F. (2009). Principal component analysis in mild and moderate Alzheimer's disease – A novel approach to clinical diagnosis, *Psychiatry Research: Neuroimaging* 173(1): 8 – 14.

Peres-Neto, P. R., Jackson, D. A. & Somers, K. M. (2005). How many principal components? stopping rules for determining the number of non-trivial axes revisited, *Computational Statistics & Data Analysis* 49(4): 974 – 997.

Petrella, J. R., Coleman, R. E. & Doraiswamy, P. M. (2003). Neuroimaging and Early Diagnosis of Alzheimer Disease: A Look to the Future, *Radiology* 226(2): 315–336.

Scarmeas, N., Habeck, C. G., Zarahn, E., Anderson, K. E., Park, A., Hilton, J., Pelton, G. H., Tabert, M. H., Honig, L. S., Moeller, J. R., Devanand, D. P. & Stern, Y. (2004).

Covariance PET patterns in early Alzheimer's disease and subjects with cognitive impairment but no dementia: utility in group discrimination and correlations with functional performance, *NeuroImage* 23(1): 35 – 45.

Selkoe, D. J. (2001). Alzheimer's disease: Genes, proteins, and therapy, *Physiological Reviews* 81(2): 741–766.

Serneels, S. & Verdonck, T. (2008). Principal component analysis for data containing outliers and missing elements, *Computational Statistics & Data Analysis* 52(3): 1712 – 1727.

Sugiura, N. (1978). Further analysis of the data by akaike' s information criterion and the finite corrections, *Communications in Statistics - Theory and Methods* 7(1): 13–26.

Tabert, M. H., Manly, J. J., Liu, X., Pelton, G. H., Rosenblum, S., Jacobs, M., Zamora, D., Goodkind, M., Bell, K., Stern, Y. & Devanand, D. P. (2006). Neuropsychological Prediction of Conversion to Alzheimer Disease in Patients With Mild Cognitive Impairment, *Arch Gen Psychiatry* 63(8): 916–924.

Van Huffel, S. (1997). *Recent Advances in Total Least Squares Techniques and Errors-in-Variables Modeling*, SIAM, Philadelphia, PA.

Wold, S., Geladi, P., Esbensen, K. & Öhman, J. (1987). Multi-way principal components-and PLS-analysis, *Journal of Chemometrics* 1(1): 41–56.

Zuendorf, G., Kerrouche, N., Herholz, K. & Baron, J.-C. (2003). Efficient principal component analysis for multivariate 3D voxel-based mapping of brain functional imaging data sets as applied to FDG-PET and normal aging, *Human Brain Mapping* 18(1): 13–21.

10

The Health Care Access Index as a Determinant of Delayed Cancer Detection Through Principal Component Analysis

Eric Belasco[1], Billy U. Philips, Jr.[2] and Gordon Gong[2]
[1]*Montana State University, Department of Agricultural Economics and Economics,*
[2]*Texas Tech University Health Sciences Center, F. Marie Hall Institute of Rural Community Health,*
USA

1. Introduction

In the past two decades, cancer mortality declined significantly in the United States (Byers, 2010). Although the reasons for the decline have not been well-established, many factors such as the reduction in the number of smokers, increased cancer screening, and better treatment may have played an important role (Byers, 2010, Richardson et al. 2010). However, disparities in cancer mortality persisted among different ethnic groups and social classes (Byers, 2010). Health status and health disparities among different social and ethnic groups are to a large degree determined by socioeconomic status and living conditions in general (Pamies and Nsiah-Kumi, 2008; World Health Organization [WHO], 2008). For example, life expectancy worldwide increased from 48 years in 1955 to 66 years in 2000 mainly as a result of improvement of overall living conditions in addition to advancement in medical science and large-scale preventive interventions (Centers for Disease Control and Prevention [CDC], 2011). Large health disparities exist between poor and rich countries or within any given rich or poor country (WHO 2008). In the case of cancer mortality due to delayed detection, socioeconomic status may determine health insurance coverage status, which in turn affects health behaviour including regular check-ups and participation in cancer surveillance among high risk groups. Regular cancer surveillance is critical for cancer control (Byers, 2010, Richardson et al. 2010). Lack of health insurance due to economic hardship may result in the delay in cancer detection.

A vexing question is how to determine socioeconomic status. Early studies associated cancer mortality with single socioeconomic indicators such as individual income, education level, below or above poverty level among others. For example, Ward et al. (2004) used the percentage of the population below the poverty line as a socioeconomic indicator and found that cancer mortality rate was 13% and 3% higher in men and women, respectively, among U.S. counties with $\geq$ 20% of the population below the poverty line as compared with those with < 10% below the poverty line from 1996 to 2000. On the other hand, Clegg et al. (2009) used education level as an indicator, and reported that lung cancer incidence was significantly higher among Americans with less than a high school education than those with a college education. Clegg et al (2009) also used family annual income as an indicator and found that lung cancer incidence rate was 70% higher in those with less than $12,500 annual income compared with those with incomes $50,000 or higher.

An alternative approach to assessing socioeconomic status is to build a composite index based on many aspects of socioeconomic status using readily available census data, which is then employed to predict health status using information from disease (e.g., cancer) registries. For example, Singh (2003) used 17 socioeconomic indicators (such as household income, median home value) derived from US census data to build a composite index for socioeconomic deprivation by factor analysis and principal component analysis (PCA). Such a composite index is believed to reflect socioeconomic status more thoroughly with multiple indictors, while PCA address the issue of inter-correlation among factors. Such a composite index tends to have a high reliability coefficient (α equal to 0.95) (Singh, 2002). The composite index is then used to predict health status. For example, Singh (Singh, 2003) reported that US mortality of all-causes was significantly and positively correlated with a deprivation index derived from US census data. Crampton et al. (1997) developed a relatively simple socioeconomic status index termed the New Zealand Index of Relative Deprivation (NZDep91) which was constructed based on nine socioeconomic variables from New Zealand census data. The NZDep91 is subsequently used to predict hospital discharge rate and all-cause mortality (Salmond et al. 1998). Albrecht and Ramasubramanian (2004) (Henceforth, A&R) modified the NZDep91 and developed a Wellbeing Index (WI) by principal component analysis using ten socioeconomic variables from US Census 2000 data. The WI is recently shown to be highly correlated with delayed cancer detection (as assessed by the ratio of late- to early-stage cases) of female genital system (FGS), lung-bronchial and all-type cancers at diagnosis among Texas counties (Philips et al., 2011).

One of the main purposes of the current study is to determine whether the percentage of late-stage cancer incidence is correlated with a newly developed index of health accessibility, which is an extension of the previously mentioned WI. We term the new index the Health Care Accessibility Index (HCAI), which is derived from principal component analysis of ten socioeconomic indicators derived from US Census Bureau's 2005-2009 American Community Survey plus two additional factors that are more closely related to health, i.e., health insurance coverage and physician supply. By examining the relationship between HCAI and late-stage cancer detection, we are able to establish whether health inequities exist in certain communities that can be related to access to health care. A high percentage of late-stage cancer cases is problematic for communities due to the often relatively low survival rates of costly procedures. The derived HCAI is compared with WI in their association with delayed cancer detection in Texas counties to determine the optimal model by the Akaike's Information Criteria (AIC) (Akaike, 1974) and Schwartz Information Criteria (SIC) (Schwartz, 1978). Another difference between WI and HCAI is that the ten socioeconomic variables for computing the HCAI includes a new variable, the median income of a county and excludes the percentage of people with disability (because of its absence in the US Census Bureau's 2005-2009 American Community Survey database).

This study also addresses several practical statistical issues regarding the choice of socioeconomic variables in association with delayed cancer detection. Firstly, A&R arbitrarily classify the WI rankings as the deciles of the first component scores retrieved from PCA. It is quite frequent for groupings to be assigned based on terciles, quartiles, or deciles, which may or may not produce a proper classification. If ad hoc metrics are used to denote an optimal number of groups to be used in any particular index, then the inference that derive from the index might not have meaning and might even add uncertainty to the representativeness of the index. For example, the grouping used for Texas might be substantially different from Delaware, given the different sizes and regional dynamics. We propose to use AIC and SIC to find the optimal number of groups and compare the goodness-of-fit between models. Secondly, it has not been statistically demonstrated that using the composite WI is superior to using each of the 10 individual socioeconomic variables in their correlation with health status such as delayed cancer detection. The optimal model of the two (a composite vs. multiple variables) will also be determined by AIC and SIC. Thirdly, we also propose the use of raw principal component scores in a regression in order to improve goodness-of-fit and compare these results to the previously mentioned models based on AIC and SIC. Using these proposed methods, we look to characterize the relationship between late-stage cancer detection with the HCAI in order to identify the existence or lack of existence of economically-rooted health inequities.

2. Review of PCA in regression analysis

2.1 Review of creating indices using PCA

Using PCA in economics is particularly appealing for applications where comparisons are warranted that comprise over a collection of variables. This method is particularly convenient and informative when the researcher is interested in the "ability" of a model to characterize a collection of variables rather than the marginal impacts between one variable and another (Greene, 2012, pg. 93). Further, the assumption of an exogenous shock is often used when evaluating marginal impacts, which are unrealistic with covariates that are highly correlated. For example, in evaluating the relative wealth or poverty in developing counties, asset indices are often built to reflect the relative wealth in order to make cross-country comparisons. For example, Booysen et al (2008) conducted a transregional survey in sub-Saharan Africa to evaluate the movement of poverty across regions over a particular time span. While "poverty" is a loosely defined term, this line of research commonly utilizes an asset index in order to evaluate the ability of citizens in each country to consume durable goods. Another example is Gwatkin et al. (2000) who use data from the Demographic and Health Survey (DHS) program to evaluate socio-economic differences between developing countries. They create a socio-economic status (SES) index, which was also used in studies such as Vyas and Kumaranayake (2006).

SES indices have also been developed in order to evaluate health outcomes in developing counties in studies such as Deaton (2003), where it is argued that health outcomes and the utilization of health services are largely different across different socioeconomic classes. Ruel and Menon (2002) create a child feeding index using responses in the DHS survey to assess the influences on child feeding practices. As in many related studies, the creation of an index is used to identify problem areas that need improved policy design or interventions.

Ewing et al. (2008) develop a sprawl index in order to evaluate the relationship between urban sprawl and health-related behaviours. As in the case of other index variables, urban sprawl can be defined only when many factors are combined. The main four factors that are include are residential density, mixture between homes, jobs, and services, strength of business centers, and accessibility of the street network. Data were from the Behavioral Risk Factor Surveillance Systems (BRFSS) from 1998 to 2000 to evaluate health activity and lifestyle. On a similar topic is a study by Pomeroy et al (1997) who evaluated perceptions and the factors that led to the appearance of a successful resource management program.

While many of these indices include continuous variables, Koenikov and Angeles (2009) and Filmer and Pritchett (2001) explore ways to incorporate discrete data into PCA. Savitry (2005) provides a comprehensive overview of all multivariate methods used in developing indices.

2.2 Review of using PC scores in regression analysis

While the developments of a reliable and robust index are quite extensive in the literature, little research has been done to evaluate the impact of incorporating an index into a regression. This is precisely because usually the index itself is the primary interest. For example, in the identification of poverty traps, one needs only an index for identification. However, if one were interested in the impact of a poverty trap on say personal liberties, then a regression would likely be needed.

There are many studies that conduct PCA to create an index that is used in a secondary regressions (Everitt and Hothorn, 2011; Everitt, 2011; Vyas and Kumaranayake, 2006). One method that is used is to rescale the principal component values. For example, Ewing et al. (2008) scale the raw principal component scores to have a mean of 100 and a standard deviation of 25. This is similar to normalizing the principal components. However, the distribution of scores might not be normal. Vyas and Kumaranayake (2006) report finding scores that are different distributions by county that include counties with normal and uniform distributions along with distributions that possess negative or positive skewness. Their results suggest that the characterization groups might be different across cases, which leads one to consider a data-driven approach. Pomeroy et al (1997) use the raw and unscaled principal component scores as the dependent variable. While marginal impacts are not easily interpreted, they can loosely be discussed directionally.

Another common approach is to use cut-off points. For example, Filmer and Pritchett (2001) split their sample into three populations based on cut-off points at the 60th and 20th percentile of the principal component scores. They define these groups to be 'low', 'medium', and 'high' socioeconomic groups. Cut-off points can also be defined a bit more arbitrarily. For example, A&R use deciles (ten groups) to define different ranges of socio-economic status. The implicit assumption made is that each group is distinctly different from the other. However, just as Vyas and Kumaranayake (2006) found differences by country it might also be possible to find that some populations need more grouping than others. Others, such as Booyens et al (2008) use quintiles in breaking up the population.

Other approaches that are more data-driven include the use of cluster analysis, which is described in some detail in Everitt et al. (2011). However, one limitation of cluster analysis as well as the previously discussed methods is that the existence of an observation in a particular group is mutually exclusive of other groups. This means that each observation must with certainty fit into a single category. However, an evolving area of research includes latent class analysis where the residency within each group is treated as an unobserved factor that is estimated with some probability, rather than absolute certainty. However, these models are also with their own set of limitations. For example, the number of classes must be defined *a priori*. Additionally, the models have been shown to be highly non-linear and can often have difficulty finding optimized values when many classes are added.

3. Data and application

3.1 Data summary and description

This study was approved by Texas Tech University Health Sciences Center Institutional Review Board with exemption for review because of its use of published data.

3.2 Cancer data

Cancer stage data from 2004 to 2008 were provided by the Texas Cancer Registry, Cancer Epidemiology and Surveillance Branch, Texas Department of State Health Services (note that cancer data from 2008 are the latest data available). This database provides cancer data by year, age, county, Hispanic origin (Hispanic vs. Non-Hispanic) as well as population size for each county. We used cancer data between 2004 and 2008 cancer data (the latest available data is the 2008 data) to match the five years of American Community Survey data although there is a one-year lag. The five most common categories of cancer are studied including breast, colorectal, FGS, lung-bronchial, and prostate cancers. Female genital system includes cervix uteri, ovary, corpus and uterus, vagina, vulva and others. We pool the five-year (2004-2008) data to calculate the numbers of age-adjusted late- and early-stage cancer cases per unit (100,000) population using 2000 USA standard population (National Cancer Institute, NCI, n.d.). We use the percentage of late-stage cases among all staged cancer cases in our analysis. The number of unstaged cancer cases is not included in the denominator because the percentage of unstaged cases varies significantly by cancer type as well as by county, and inclusion of such cases in the denominator would result in uncertainty in estimating the percentage of late-stage cancer cases. Carcinoma in situ and localized cancers are considered as early-stage while cancers defined as "regional, direct extension only", "regional, regional lymph nodes only", "regional, direct extension and regional lymph nodes", "regional, NOS" and "distant" are considered as late-stage (Philips et al. 2011).

3.3 Socioeconomic status data

Socioeconomic status data are derived from the U.S. Census Bureau's (n.d.) 2005-2009 American Community Survey. Since this survey does not provide percentage of people with disability, the present study uses the remaining nine of the 10 socioeconomic variables originally used to build the Wellbeing Index (WI) developed by A&R and are listed in Table 1. We add median income (from the 2005-2009 American Community Survey) so that the total number of socioeconomic variables is still ten in the current study.

3.4 Data of factors more closely related to health

Data for the percentage of uninsured and percentage of obese individuals are obtained from Texas State Data Center (n.d.). The number of physicians and estimated population size in each county from 2004 to 2008 are derived from Texas Department of State Human Services (DSHS, n.d.). Physician supply is the number of physicians per 1,000 residents in each county. Physicians considered are those with medical doctor (MD) and/or doctor of osteopathy (DO) degrees who worked directly with patients. Residents and fellows; teachers; administrators; researchers; and those who were working for the federal government, military, retired, or not in practice were excluded from the total of physicians by DSHS (n.d.).

	Description	Weighted Mean	Std Error	Min	Max
% Single Parent	% people in single parent households	18.89	0.54	0.00	27.55
% No High School	% people over 18 without high school	21.38	1.17	2.60	53.59
% Unemployed	% people unemployed	6.87	0.15	0.00	21.58
% Income Support	% people with income support	28.20	1.47	0.00	66.64
% Below Poverty	% people in households below poverty level	15.90	0.88	0.00	46.81
% No House	% people not living in own home	34.19	1.74	11.88	70.83
% Few Room	% people living in homes with too few bedrooms	4.97	0.47	0.00	15.79
% No Phone	% people in households without phone	5.33	0.20	0.00	18.23
%No Car	% people in households without car	6.55	0.40	0.00	17.55
Median Income	Median Income (in $1,000s)	49.63	1.73	20.38	82.55
%Uninsured	% people without health insurance	24.47	1.07	14.20	38.10
Physician supply	Number of physicians per 1,000 residents	0.16	0.01	0.00	0.33
% Obese	% people with BMI above 30	28.90	0.33	23.80	32.80
% Hispanic	% Hispanic	35.60	3.19	2.44	97.15
Population	Population (in 1,000s)	636.02	218.75	0.02	1,912

Table 1. Description and weighted summary statistics of relevant covariates

4. Statistical analysis and results

4.1 Computing the health care accessibility index

While access to health care may be determined by socioeconomic status in general, health insurance coverage and health care services (number of physicians and or hospitals relative to local population) may more directly impact on health as discussed above. In this study we add the latter two variables to the 10 socioeconomic indicators for a principal component analysis to build Health Care Accessibility Index (HCAI).

There are two well-known benefits to creating an index rather than including all of the components in the index into a regression. First, the index itself can serve as a variable of interest to identify (in our study) areas of extreme health care obstacles. Second, the index can significantly reduce the degrees of freedom in a regression while preserving information from the variables. In this study we also find a third benefit, which is that if we are careful about how we use the index in regression analysis, we can also improve goodness-of-fit.

While the A&R study presents an appealing start to our research, it lacks two important components. First, it only includes socioeconomic variables that are intended to provide an index for wellbeing or economic deprivation. While their index provides a good basis to evaluate the ability to pay for health services, it lacks health-specific variables. Second, the study presents a less than appealing method for incorporating the index into regression by ranking the first principal component scores into deciles. This essentially groups the observations into 10 different bins which, may or may not be the optimal number of bins to use in grouping regions. To fix the second issue of rank ordering, we use AIC and SIC to determine the optimal number of bins (more discussion on this follows in the next subsection).

In order to fix the first issue, we add three components which are essential to access to health care services which include median income, the percentage of the uninsured, and the number of physicians per 1,000 residents. All variables are computed at the county-level. These variables are intended to account for the access residents within a county have to health care services. Lack of health care insurance in clearly a hurdle in obtaining affordable health care services and is another factor included in this analysis. Finally, the number of physicians allows for an evaluation into geographic access to health care services. Does the county in question have an adequate medical infrastructure to prevent and detect illness when it arises? Some Counties in Texas (particularly in the western region) are geographically isolated. If adequate care is not geographically close to residents, the economic cost to receiving care increases in terms of time off work and travel costs.

In compiling the twelve variables, we are able to derive an index for access to health care services through the use of PCA. While the goal is to consolidate a group of many variables into a smaller set of linearly related variables (principal components), it is often the case that multiple principal components are needed to explain a substantial proportion of the variation in the independent variables. Results from the initial PCA is shown below in table 2.

The first principal component explains 41% of all variation in the variables included in the index. The influence is also spread across relatively evenly across all included variables, with the exception of direct patient physician supply (DPC). The principal component scores from this regression are used as the HCAI. Notice the negative score associated with median income, which is consistent with a negative relationship between income and obstacles to receiving health care services. The second principal component explains an additional 16% of the variation in the variables and is largely influenced by physician supply, percent uninsured, percent without their own house, and percent of single parents.

These results provide us with a couple of items. First, we have the first five principal components, which can be used rather than the original 13 variables in order to shrink the necessary variables while still preserving almost all of the variation in the variables. One notable and helpful point in this analysis is that each component is orthogonal to the others,

| | Principal Component Score | | | | |
Variable	1	2	3	4	5
% Single Parent	0.31	0.22	0.22	0.01	0.21
% No High School	0.36	-0.13	-0.36	-0.03	0.15
% Unemployed	0.24	0.05	0.31	0.71	-0.43
% Income Support	0.18	-0.56	0.23	-0.05	0.28
% Below Poverty	0.39	-0.08	0.09	0.00	-0.02
% No House	0.18	0.45	0.21	-0.30	-0.24
% Few Room	0.27	0.17	-0.48	0.24	0.17
% No Phone	0.28	0.09	-0.04	-0.50	-0.48
%No Car	0.36	-0.03	0.22	0.12	0.05
Median Income	-0.35	0.31	-0.18	0.26	-0.08
%Uninsured	0.31	0.24	-0.41	0.07	0.11
DPC	-0.01	0.47	0.37	-0.04	0.57
Eigenvalue	4.89	1.94	1.36	0.76	0.74
Difference	2.95	0.58	0.60	0.02	0.11
Proportion	0.41	0.16	0.11	0.06	0.06
Cumulative	0.41	0.57	0.68	0.75	0.81

Table 2. PCA results from variables in health access index

meaning the independent variables will be uncorrelated and avoids the issue of multicollinearlity which arises from including all 13 variables into the regression. A notable problem when multicollinearity is particularly acute is that it is difficult to isolate marginal relationships between competing variables, which often leads to high standard errors. Second, the first principal component scores can be used to develop the index of interest in the following way. The index is created by recognizing that the first principal component can be written as a linear combination of the original variables such that

$$Prin1 = a_{1,1}x_1 + a_{1,2}x_2 + \cdots + a_{1,13}x_{13} \tag{1}$$

where $a_1 = (a_{1,1}, a_{1,2}, \ldots, a_{1,13})$ include the parameter estimates in table 2 and $x = (x_1, x_2, \ldots, x_{13})$ include the 13 parameters used in the index from table 1. $Prin1$ is then computed for each observation and can be ranked to present a ordering of health accessibility. The second component is then derived based on the remaining variability and results in a_2 which leads to $Prin2$. The second component is derived based on the restriction that $a_2'a_1 = 0$ so that $Corr(Prin1, Prin2) = 0$. Additional components can be derived in the same fashion so that they are uncorrelated with all prior components.

The rankings associated with the first principal component, $Prin1$, is used as the basis of our HCAI. To illustrate the regional dynamics of this new index, we present figure 1 below.

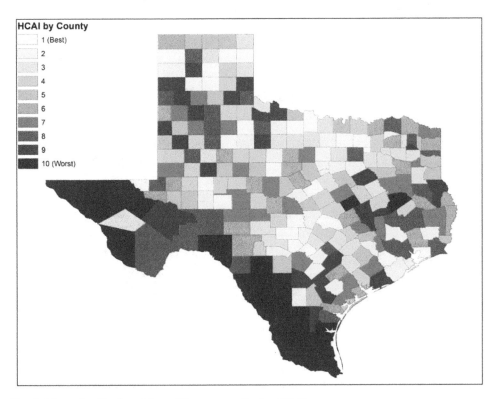

Fig. 1. Map of ordinal ranking of Texas counties by HCAI

Figure 1 shows that the counties that have poor health care access are focused along the Mexican border. These counties tend to have a large percentage of Hispanics in the population. Other counties further north are also found to have high HCAI rankings. Some of the better HCAI rankings are in the more urban areas of Texas that are near the largest metropolitan areas of Dallas, Fort Worth, Houston, San Antonio, and Austin. The large size of Texas, as well as its diversity in the variables used in this study, provides an opportunity to evaluate the how these factors influence late-stage cancer detection.

In the present study, the percentage of late stage cancer cases is hypothesized to correlate with the developed HCAI. Figure 2 below shows a scatter plot of the dependent variables along with the first component scores of HCAI for each cancer type that is evaluated in this study. Given the principal component scores shown in table 2, it is clear that a county that has a low degree of health access will possess a high component score. These same areas with a low degree of health access correspond to high rankings. Given this, it is not surprising that some of the slopes shown in figure 2 are significantly positive. Lung cancer is one example where the positive slope is particularly striking. On the other hand, the relationship with the HCAI is shown to be positive but relatively flat for colon and prostate cancers. Given the different speeds of progression in different cancers, it should not be surprising that the relationships differ across cancer types.

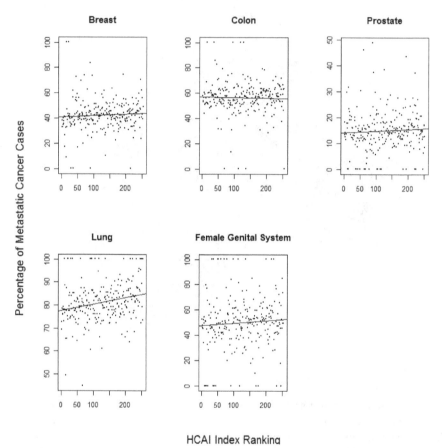

HCAI Index Ranking

Fig. 2. Plot of the percentage of late-stage cancer cases with HCAI first principal component score, by cancer type

4.2 Weighted tobit regression model

The dependent variable of interest used in this analysis is the log of the sum between one and the percentage of late-stage cancer cases among all staged cancer cases. Some counties experienced no late-stage cancer cases and other experiences all late-stage cancer cases. These are all from relatively small counties. However, this does mean that censoring is an issue that will need to be dealt with in this study. As shown in table 3, there is some degree of censoring for all regressions in our analysis. For example, the percentage of late stage breast cancer cases was lower censored at 0 in four cases and upper censored in two cases. In the lower censoring case, the county experienced no late-stage cancer cases but experienced only early-stage cases. Alternatively, some counties experienced only late-stage cases resulting in an upper bound of 100%. This double-bounded censoring occurred for all cancer types to varying degrees.

If censoring is not accounted for and usual least squares methods are used, we are assured of biased estimates (Amemiya, 1973). So, in order to obtain unbiased parameter estimates, we use the Tobit model (Tobin, 1958) to correct for censoring. The Tobit model assumes the dependent variable, Y, is a partially observed derivative of a latent variable, Y^*. When censoring occurs at zero, the two are related with the following relationship, $Y = \max(0, Y^*)$. In this particular application of a double-bounded Tobit model, we assume a latent variable, Y^*, with the following relationship to the observed variable, Y.

$$Y = \begin{cases} 0 & if \ Y^* < 0 \\ Y^* & if \ 0 \leq Y^* \leq 100 \\ 100 & if \ Y^* > 100 \end{cases} \tag{2}$$

A more comprehensive discussion on the Tobit model can be found in Amemiya (1974). Residuals are weighted by population size in order to account for the wide variety in the size of counties. Weighted statistics provide a more accurate representation of the population of interest, which is the state of Texas in this application.

Type	Weighted Mean	Std Error	Min	Max	Limit obs.	
					Lower	Upper
Breast	40.46	0.48	0.00	100.00	4	2
Colon	57.92	0.81	0.00	100.00	4	5
Prostate	14.45	0.41	0.00	48.87	18	0
Lung	79.28	0.59	44.98	100.00	0	20
FGS	50.69	0.55	0.00	100.00	13	13

Table 3. Percentage of late-stage cancer cases, by cancer type

4.3 Incorporating the index into regression analysis

While the selection of principal components is well established and can be found in most intermediate econometrics text books, the usage of that information in a regression is less understood. For this reason, we use five separate regressions for each cancer type to compare different methodologies that could be used to integrate an index or information in an index within a regression context. We use AIC and SIC to compare model fit and include two additional variables that are hypothesized to influence the percentage of late stage cancer cases. These variables include the log of the percentage of obese individuals and the log of percentage of Hispanic individuals. The five models of interest include the following specifications:

Model 1. The composite ordinal WI (from A&R) based on deciles according to the first principal component scores

Model 2. All 12 variables in the composite HCAI

Model 3. The raw scores of the first two principal components using variables in HCAI

Model 4. The composite ordinal HCAI based on grouping of the first principal component scores that minimizes AIC

Model 5. The composite ordinal HCAI based on grouping of the first principal component scores that minimizes SIC

The first method uses the method proposed by A&R, which assumes the use of deciles to group the ordered WI. This model provides a prior baseline under which to evaluate all of the proposed models. As A&R point out, once the index is used to define groups, those groups are compared in an ordinal manner, implying the se of ordinal and not continuous variables in a regression. Model 2 does not make use of PCA and places all the variables used in the HCAI directly into the regression. This model also provides another baseline model in that we can compare the methods used in PCA to that of this model which simply uses the variables without any PCA. In order for PCA to be effective in our analysis, it will need to demonstrate an improvement over model 2. In our original analysis we included two additional regressions that were omitted from our final analysis because they were not found to improve upon the utilized models. These models included (1) the raw WI variables and (2) decile grouping from the HCAI. Using raw WI variables was consistently outperformed by model 2, while the decile grouping of HCAI was consistently outperformed by models 4 and/or 5 unless they selected 10 groups (as occurred in a couple of instances).

The final three models make use of the PCA results based on the HCAI. These are the models of interest in the sense that we hypothesize that they will improve upon Models 1 and 2. Model 3 uses the raw scores of first two principal components in the regression. The scores of first two components are used since they minimize both AIC and SIC in all of the used regressions. Component scores are fitted values using the parameter values listed in table 2 and the associated variables for each observation. Therefore, each observation will contain a unique component score. The first two principal component scores explain 57% of the total variation in the used variables based on the results in table 2. While it might seem more appropriate to include say the first five principal components in order to explain 81% of the total variation, the final three components tend to be significant are the model does not fit the data as well (in terms of AIC and SIC) as when only the first two scores are included. This model provides a set of continuous variables that can be used to determine late-stage cancer detection. Models 4 and 5 use AIC and SIC to find the optimal number of groups to be used for HCAI. This search method is conducted by running all models up to 40 groups in order to find the single model that minimizes AIC or SIC. Observations are evenly divided across the selected number of groups which are used as categorical variables in our analysis.

These five models are then compared by using AIC and SIC metrics to assess goodness-of-fit. In order to declare one of the proposed models as an improvement, it will need to have values of both AIC and SIC that are lower than models 1 and 2. Since maximum likelihood methods are commonly used to estimate Tobit estimates, an appealing goodness-of-fit measure is that of AIC and/or SIC since they are both easily derived from maximum likelihood outputs. The two can be expressed as follows

$$AIC = 2k - 2LL \qquad (3)$$

$$SIC = klog(k) - 2LL \tag{4}$$

where k is the number of parameters and LL is the maximized log-likelihood value that corresponds with the optimal parameter estimates. Both information criteria measure are composed of two parts, which include two times LL and the penalty for adding new parameters. For models that contain more than 8 parameters ($k > 8$), SIC provides a heavier penalty for adding new variables. For this reason, AIC tends to support more overfit models while SIC tends to support more underfit models. Both AIC and SIC provide a basis upon which to select from nested models by minimizing the associated criteria. However, Koehler and Murphree (1988) point out that in their analysis the two criteria indicate different models 27% of the time. Their research suggests that SIC is often preferred due to the improved predication accuracy from SIC-preferred models. However, as pointed out by Kuha (2004), the preference for AIC or SIC can often rely on the data generation process and when AIC and SIC both select the same model, it is shown to be a more robust result. With these points in mind, we proceed by using AIC and SIC criteria separately to determine the appropriate model and determine whether our proposed models provide an improvement on past models.

A&R suggest that when using their index of socioeconomic deprivation, one should rank the first principal component then group this ranking into deciles, so that groupings will occur at 10% intervals. For example, if a county receives a grouping of 1, it is found to be in the lower 10% with regard to the index. We find this ranking method to be arbitrary and therefore suggest the use of AIC/SIC to determine the optimal number of deciles. This is achieved by re-running each model with the number of groups changing every time starting with 1 group (or rather no grouping at all) and up to 40 groups. Forty groups were selected because AIC and SIC did not show significant improvement after that point. With such a large and diverse state as Texas, it seems that the same rank grouping would be substantially different than for some smaller state. Of importance for this study is the generalizeability of our study to other areas. For example, does the method that we propose provide a general enough method on which to evaluate any region? While the selection of a specific grouping may work best in one situation, a more data-driven approach (such as the one we propose) will provide more flexibility in evaluating different situations.

Also included in each model will be two additional variables in order to evaluate the impact of different specifications on parameter estimates, since typically marginal inference is an important component of any analysis (particularly in economics). These variables include the log of percentage of Hispanic and the log of percentage classified as obese. Both variables are likely to have some degree of correlation with the index ranking but to varying degrees. This provides a look at two variables that have varying relationships with the index in question.

4.4 Empirical results

4.4.1 Model selection

The first tasks in the empirical section of this analysis is to determine grouping based on the candidate model that minimizes AIC and SIC (models 4 and 5) for all 5 cancer types models (as outlined in section 3.2). The selected numbers of groups are shown in table 4.

	AIC	SIC
Breast	11	5
Colon	10	1
Prostate	9	1
Lung	34	8
FGS	1	1

Table 4. Number of groups determined based on minimum AIC and SIC criteria for models 4 and 5

The variability in the selected number of groups, suggests that arbitrarily extracting a grouping number might not be optimal. For example, in our analysis, lung cancer makes use of many groups while FGS cancer suggests only 1 group (which means no group distinctions at all). To illustrate the meaning of a group, the use of 34 groups implies there are 34 groups (each with 7-8 observations). In the regression, this implies that 33 new classification variables are added which are evaluated relative to the omitted group. Each model is run using the weighted Tobit model to assess the goodness-of-fit as well as parameter estimates of variables outside of the proposed index for all 5 cancer-types as well as with all cancer types. The regression results are below in table 5.

Table 5 shows the five models used in this study along the column headers, while row headers provide space for AIC and SIC measures for each cancer type. The purpose of this analysis is to evaluate the newly developed models (3-5) and compare them to two baseline models (1-2). Recall that in order for models 3-5 to unambiguously improve the baseline models 1-2, they must improve both AIC and SIC measures. Each model that improves upon the baseline models is shaded in grey in table 5.

For each cancer type, at least one of the three proposed models improved over the baseline models. Additionally, the use of a data-driven method such as AIC or SIC provides a clear improvement over the baseline models in each cancer-type. It is interesting to note that the best fitting model is not unambiguously from using AIC or SIC. This is a common problem that has been discussed in other studies such as Koehler and Murphree (1988). The only instance when AIC does not improve upon the base models is in the case of lung cancer, which suggests the use of 34 groups, which is unusually high based on the other selections. Alternatively, SIC provides a heavier penalty for new variables and tends to support lower parameterized models. This is clearly the case when SIC suggests the use of 1 group 60% of the time, where no group distinctions are provided. In two out of those three instances, SIC provided a clear improvement over baseline models. The use of raw principal component scores are also considered in model 3 and are shown to clearly improve on the baseline models in breast and FGS cancer types. In the case of breast cancer, it is interesting to note that this model provided the lowest AIC and SIC, making it a robust selection for best model fit.

The results from regressions on colon cancer provide some interesting insights. First, AIC suggests the use of deciles, which is in line with A&R. While this research makes the argument that deciles will not always be satisfactory, it can be optimal in some situations. It is also worth noting that model 4 improves in both AIC and SIC over model 1. These two models are identical with the exception of the new index that is used in model 4, which indicates its improved predictive power over the WI. The model suggested by SIC is to include only 1 group, which are really no group distinctions at all. This also presents the possibility that when an index might not always provide valuable information for a model, which can only be detected through the use of a data-driven method, such as information criteria. However, model 5 is not the favored model in this scenario as it does not improve upon the earlier models. For this reason, model 4 is the preferred model for colon cancer.

| | | WI | HCAI | | | |
| | | Deciles | All Vars | Prin Scores | Optimal AIC | Optimal SIC |
Type	Model	1	2	3	4	5
Breast	AIC	-384.601	-389.795	-404.036	-392.419	-383.926
Breast	SIC	-338.822	-333.451	-382.907	-343.118	-355.755
Colon	AIC	-397.097	-405.139	-395.164	-408.757	-392.295
Colon	SIC	-351.266	-348.732	-374.011	-362.926	-378.193
Prostate	AIC	-24.605	-23.846	-25.446	-34.810	-28.660
Prostate	SIC	21.174	32.498	-4.318	7.447	-14.574
Lung	AIC	-859.488	-855.730	-851.945	-878.167	-869.566
Lung	SIC	-813.554	-799.195	-830.744	-747.431	-830.699
FGS	AIC	-132.891	-131.983	-144.948	-143.995	-143.995
FGS	SIC	-87.269	-75.833	-123.891	-129.958	-129.958

Table 5. Weighted Tobit goodness-of-fit regression results from 5 alternative models, by cancer type (Dependent variable: logged sum of one plus percentage late-stage cancer cases)

Based on figure 3, lung cancer is one particular type of cancer that is expected to be highly correlated with the HCAI. Because of the high degree of positive correlation, it is not surprising that AIC and SIC both suggest relatively high grouping values. For example, AIC suggests using 34 groups, while SIC suggests the use of 8 groups. However, while the AIC selected model has a very low AIC value (-878.167), SIC is not as impressive given the relatively large penalty factor for each of the 34 groups. Thus, the SIC selected group is able to improve over the base models in terms of AIC and SIC and is therefore the preferred model for lung cancer.

Many variables under study are inter-correlated. For example, figure 3 shows that percentage of the uninsured in Hispanic is significantly higher than that in non-Hispanic populations in Texas. Figure 4 shows that the percentage of late-stage cancer cases in Hispanics is higher than non -Hispanics in Texas. Hispanics also tend to have higher percentage obese individuals and socioeconomically deprived individuals. These facts suggest the necessity to adjust for covariates in assessing the association between the HCAI and delayed cancer detection. This will also allow determination of the role of ethnicity in the delayed cancer diagnosis.

We acknowledge in this research that not all types of cancer will have a significant relationship with the HCAI index. While the types of cancers that do have a correlation show health inequities, not all cancer types are thought to have inequities. In particular, we expect for cancers where detection is at a high cost (such as breast, colon, and lung cancer) to be particularly susceptible to health inequities. For example, in order to detect lung colon cancer, a costly colonoscopy is necessary which will have a lower compliance rate in individuals with poor access to health care services.

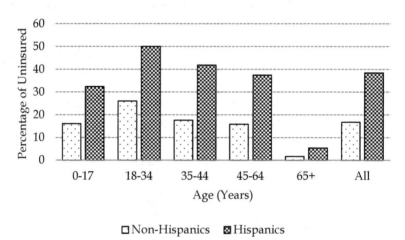

Fig. 3. Percentage of the Uninsured in Hispanic and Non-Hispanic populations in Texas.

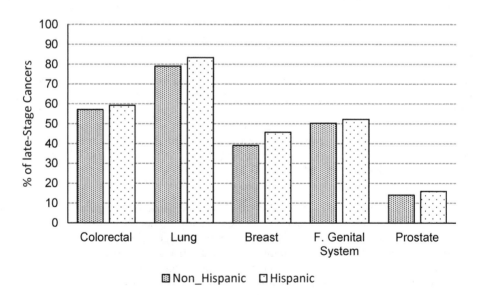

Fig. 4. Percentage of late-stage cancer cases in Hispanics and non -Hispanics in Texas.

4.4.2 Regression results

Below are included the parameter estimates resulting from the optimally selected model based on the criteria stated above. In all four cases, at least one of the newly proposed models outperformed the baseline models in terms of AIC and SIC. In the case of breast cancer, late-stage cancer detection was best determined using the raw principal component scores. Parameter estimate results are listed in table 6.

Variables	Estimate	SE	p-value
Intercept	2.800	0.691	<.0001
log(% Hisp)	-0.039	0.019	0.040
log(% Obese)	0.325	0.218	0.136
Prin1	0.024	0.005	<.0001
Prin2	-0.021	0.006	0.001
Sigma	0.106	0.005	<.0001

Table 6. Weighted Tobit regression results of parameter estimates of the % late-stage cases for breast cancer, using raw principal component scores from composite HCAI (model 3).

Based on the parameter estimates in table 2, it is clear that a high principal component score is associated with lower health care access. For this reason, it's not surprising to see a positive and significant parameter estimate of the %late-stage breast cancer cases on *Prin1*. However, *Prin2* is negatively associated with % late-stage breast cancer cases. Percent Hispanic appears to have a negative impact on late-stage cancer detection. As shown in Figure 3, Hispanics, as a group, actually have a higher percentage of late-stage cancer cases relative to non-Hispanics. However, once we control for obesity and access to health care, we see that a 10% increase in the % Hispanics in a population is associated with a 0.39% decrease in late-stage breast cancer cases. As shown in figure 3, the percentage of uninsured (which is included in HCAI) is much higher in Hispanic populations than non-Hispanic populations. This speaks to the high degree of correlation between HCAI and % Hispanic variables. Obesity appears to have an insignificant impact, although it is worth noting that obesity is highly correlated with % Hispanic as well as the HCAI.

Table 7 shows the results for late-stage colon cancer incidents, where the use of deciles based on model 4 was selected based on its low AIC and SIC. Paradoxically, counties with the worst degree of access to health care (rank9) have a lower percentage of late-stage cancer cases. After detection of adenomatous polyps (not carcinoma yet) by screening, polypectomy is generally performed before the polyps evolve to early-stage cancer (Philips et al., 2011). Since counties with better health care accessibility tend to have higher CRC screening rate, many would-be early stage CRC cases are eliminated by polypectomy, resulting in a reduction in early-stage CRC cases. This may explain the paradoxical negative association between the percentage of late-stage CRC cases and HCAI. In order for the HCAI to be more useful in finding areas that are problematic in late-stage colon cancer determination, the data necessary needs to include the finding of colon cancer precursors, which is not present in our data.

Variables	Estimate	SE	p-value
Intercept	2.592	0.710	0.001
log(% Hisp)	0.026	0.020	0.182
log(% Obese)	0.390	0.220	0.076
rank1	0.106	0.041	0.009
rank2	0.119	0.043	0.006
rank3	0.104	0.038	0.006
rank4	0.021	0.039	0.582
rank5	0.079	0.051	0.120
rank6	0.076	0.047	0.106
rank7	0.053	0.032	0.092
rank8	0.121	0.027	<.0001
rank9	0.062	0.047	0.189
Sigma	0.102	0.005	<.0001

Table 7. Weighted Tobit regression results of % late-stage colon cancer cases, composite HCAI grouping based on 10 groups (model 4).

The parameter estimate results for table 8 are shown below for prostate cancer. Almost all of the included parameter estimates are insignificant, which is consistent with previous results likely for the same reason (Philips et al., 2011).

Variables	Estimate	SE	p-value
Intercept	2.395	1.420	0.092
log(% Hisp)	0.014	0.041	0.731
log(% Obese)	0.077	0.443	0.862
rank1	-0.009	0.080	0.916
rank2	-0.005	0.089	0.964
rank3	-0.046	0.074	0.538
rank4	0.054	0.084	0.522
rank5	0.095	0.107	0.374
rank6	0.044	0.064	0.495
rank7	0.139	0.062	0.024
rank8	-0.056	0.058	0.338
Sigma	0.214	0.010	<.0001

Table 8. Weighted Tobit regression results of % late-stage cases for prostate cancer, composite HCAI grouping based on 9 groups (model 4).

Table 9 shows the results for late-stage lung cancer detection. Based on figure 2, the relationship between lung cancer and HCAI has the strongest correlation. Based on this, it is not surprising to see so many significant variables in the results. For example, groups 1, 2, and 6 are statistically lower than the reference group 8 (omitted). Relative to the other groups, the omitted group (8) has the worst access to health care. It is worth mentioning that as the ranking group increases, the estimate decreases in general. Given that the HCAI also captures socioeconomic factors, it is also likely that counties with high HCAI have higher smoking rates than other counties given the relationship between income and the prevalence of smoking (Laaksonen et al., 2005).

Variables	Estimate	SE	p-value
Intercept	4.469	0.275	<.0001
log(% Hisp)	0.002	0.008	0.894
log(% Obese)	-0.018	0.085	0.836
rank1	-0.073	0.016	<.0001
rank2	-0.019	0.017	0.266
rank3	-0.048	0.015	0.001
rank4	-0.025	0.016	0.115
rank5	-0.020	0.017	0.226
rank6	-0.040	0.011	0.001
rank7	0.004	0.011	0.777
Sigma	0.042	0.002	<.0001

Table 9. Weighted Tobit regression results of estimate of % late-stage cases for lung cancer, composite HCAI grouping based on 8 groups (model 5).

Finally, table 10 shows the parameter estimates associated with late-stage detection for FGS cancer. While the optimal model does not contain any information for the HCAI, the results show significant relationships with the percentage of Hispanics (a negative association) and obese (a positive association) individuals. The latter finding is consistent with previous report that obesity is a risk factor for endometrial cancer (a type of FGS cancer), while the former again suggests that the higher percentage of late-stage FGS cancer in Hispanics (Fig. 3) is due to their lack of access to health care and or socioeconomic deprivation.

Variables	Estimate	SE	p-value
Intercept	0.996	0.910	0.274
log(% Hisp)	-0.080	0.031	0.010
log(% Obese)	0.955	0.296	0.001
Sigma	0.177	0.008	<.0001

Table 10. Weighted Tobit regression results for female genital system (FGS) cancer with no HCAI grouping (models 4 and 5).

5. Final discussion

These results demonstrated in the previous section suggest that data-driven methods such as AIC and SIC as well as the use of raw principal component scores show significant improvement in terms of model-fit when explaining the percentage of late stage cancer cases. Additionally, the newly developed HCAI improves upon the index derived by A&R when evaluating late stage cancer detection. These findings add insight into future studies that will hopefully more effectively utilize PCA in different applications.

The use of data-driven techniques clearly has the ability to provide more flexibility in using PCA in a variety of applications without using rigid rules for grouping ordered indices. While some indices may be appropriately developed, the usage of this information in a regression or grouping context still desires some advances. Using this methodology, we found that the HCAI is positively correlated with the percentage of late-stage cases for breast and lung cancers. This research identifies that in order to promote early cancer detection policy makers should increase the rate of health access as defined in the HCAI. Additionally, the relative lack of explanatory power for the other cancers in this study points to some necessary areas for future research. First, in the case of colon cancer, more detailed data are needed. In order to properly assess colon cancer, data are needed that take account for pre-cancer detection of polyps and adenomas. Without that information, late- and early-stage cancer cannot be properly identified because early detection in the case of colon cancer is detection before it is cancerous. For prostate and FGS cancers, the index was not very explanatory which indicates a lack of health inequality in these areas that can be traced back to health access. Without a doubt, each cancer type is different and in this research we have identified a few types of cancers (breast and lung primarily) where health inequalities exist. These cancers tend to have relatively high costs associated with detection.

The issue of creating an index of health accessibility to explain disparities in health outcomes is an area of much importance in developing and developed counties. The present study shows that health care accessibility as measured by the HCAI impacts delayed detection of several cancers consistent with the results based on data in 2000 (Philips et al., 2011). The positive correlations between the two variables are statistically significant or marginal for all these cancers studied after controlling for several other potential determinants. This finding suggests that socioeconomic deprivation, health insurance coverage and health care service significantly impacts delayed detection of these cancers independent of percentage of Hispanics or percentage of obese individuals in counties of Texas which is one the states with most severe physician shortage and lowest health insurance coverage. Thus, we not only have for the first time proposed the HCAI but also for the first time validated its utility in its correlation with delayed cancer detection.

Previous study showed that WI was significantly associated with delayed detection (assessed by the ratio of late- to early-stage cancer cases similar to the percentage of late-stage cases) of FGS and lung-bronchial cancers but not breast cancer (Philips et al., 2011) in contrast to the results of the current study. The difference in the results is due to difference in study design with which several covariates are entered in the regression in the current study. Particularly, the HCAI covers not only socioeconomic variables but health insurance and health service (physician supply) as well.

It is of interest to note that Hispanics have higher percentage of late-stage cancer cases and also have higher percentages of obese individuals, higher percentages of uninsured individuals. In our multiple regression analysis, we find that delayed cancer detection is actually negatively associated with Hispanic after adjusting for socioeconomic status, physician supply, and percentage of uninsured, which are all included in the HCAI, and percentage of obese individuals. This suggests that their delay in cancer diagnosis is likely due to these factors rather than Hispanic culture per se or genetic predisposition.

These findings provide the evidence-base critical for decision makers to establish policies to promote early detection for effective cancer control targeting specific berries such as physician shortage, lack of health insurance and to improve socioeconomic conditions in general.

One promising avenue that was not undertaken in this study is that of latent class models in grouping. Latent class models provide a couple of advantages over the methods used here: (1) class identification is treated as an unknown variable, which is different from the mutually exclusive technique use in most studies; and (2) classes need not be similar sizes. One major limitation of finite mixture models is that they have been shown to be highly non-linear when more than a few groups are used. In our particular application, this would be a major obstacle given the high number of classes suggested by some models.

6. Acknowledgment

The authors wish to acknowledge the assistance from Kris Hargrave. Cancer incidence data have been provided by the Texas Cancer Registry, Cancer Epidemiology and Surveillance Branch, Texas Department of State Health Services, 1100 W. 49th Street, Austin, Texas, 78756, http://www.dshs.state.tx.us/tcr/default.shtm, or (512) 458-7523.

7. References

Abseyasekera, S. (2005). Multivariate Models for Index Construction, In: *Household Surveys in Developing and Transition Counties: Design, Implementation, and Analysis*, pp. 367-388, United Nations, 92-1-161481-3, New York, NY, USA.

Albrecht, J. and Ramasbramanian, L. (2004). The Moving Target: A Geographic Index of Relative Wellbeing, *Journal of Medical Systems*, Vol.28, No.4, pp. 371-384, 1573-589X

Amemiya, T. (1973). Regression Analysis when the Dependent Variable is Truncated Normal, *Econometrica*, Vol.41, No.6, pp. 997-1016, 1468-0262

Amemiya, T. (1984). Tobit Models: A Survey, *Journal of Econometrics*, Vol.24, No.1-2, pp. 3-61, 0304-4076

Akaike, H. (1974). A New Look at the Statistical Identification Model, *IEEEE Transactions on Automatic Control*, Vol.19, No.6, pp. 716-723, 0018-9286

Berenger, V. and Verdier-Chouchane, A. (2007). Multidimensional Measures of Well-Being: Standard of Living and Quality of Life Across Counties, *World Development*, Vol, 35, No. 7, 1259-1276.

Booysen, F., Van Der Berg, S., Burger, R., Von Maltitz, M. (2008). Using an Asset Index to Assess Trends in Poverty in Seven Sub-Saharan African Countries, *World Development*, Vol. 36, No. 6, 1113-1130.

Byers T. Two decades of declining cancer mortality: progress with disparity. Annu Rev Public Health 2010, 31:121-32.

Centers for Disease Control and Prevention (CDC). Ten great public health achievements--worldwide, 2001-2010. MMWR Morb Mortal Wkly Rep. 2011;60:814-818

Clegg LX, Reichman ME, Miller BA, Hankey BF, Singh GK, Lin YD, Goodman MT, Lynch CF, Schwartz SM, Chen VW, Bernstein L, Gomez SL, Graff JJ, Lin CC, Johnson NJ, Edwards BK. Impact of socioeconomic status on cancer incidence and stage at diagnosis: selected findings from the surveillance, epidemiology, and end results: National Longitudinal Mortality Study. Cancer Causes Control. 2009;20:417-435.

Cortinovis, I., Vella, V., and Nkiku, J. (1993). Construction of a Socio-Economic Index to Facilitate Analysis of Health Data in Developing Countries, *Social Science & Medicine*, Vol, 36, No. 8, 1087-1097, 0277-9536.

Crampton P, Salmond C, Sutton F. NZDep91: a new index of deprivation. Soc Policy J New Zealand. 1997;9:186-193

Davis, B. (2003). *Choosing a Method for Poverty Mapping, Food and Agriculture Organization of the United Nations*, 92-5-104920-3, Rome, Italy. Accessed at http://www.fao.org/DOCREP/005/y4597e/y4597e00.HT on Oct. 7, 2011.

Deaton, A. (2003). Health, Inequality, and Economic Develoment. *Journal of Economic Literature*, Vol 41, 113-158,

Everitt, B. and Hothorn, T. (2011). *An Introduction to Applied Multivariate Analysis with R*, Springer Science, 978-1-4419-9649-7, New York, NY, USA.

Everitt, B., Landau, S., Leese, M. and Stahl, D. (2011). *Cluster Analysis* (5th ed.), John Wiley & Sons, 978-0470220436, Chichester, UK.

Ewing, R., Schmid, T., Killingsworth, R., Zlot, A., Raudenbush, S. (2003). Relationship Between Urban Sprawl and Physical Activity, Obesity, and Morbidity. *American Journal of Health Promotion*, Vol. 18, No. 1, 47-57.

Filmer, D. & Pritchett, L.H. (2001). Estimating Wealth Effects without Expenditure Data – or Tears: An Application to Educational Enrollments in States of India, *Demography*, Vol, 38, No. 1, 115-132.

Greene, W. (2012). *Econometric Analysis* (7th edition), Pearson Education, Inc., 0-13-139538-6, Upper Saddle River, NJ, USA.

Gwatkin, D.R., Rutstein, S., Johnson, K., Pande, R., and Wagstaff, A. (2000). Socio-Economic Differences in Health, Nutrition, and Population. HNP/Poverty Thematic Group, World Bank, Washington, DC.

Koehler, A. & Murphree, E. (1988). A Comparison of the Akaike and Schwarz Criteria for Selecting Model Order, *Journal of the royal Statistical Society. Series C (Applied Statistics)*, Vol.37, No.2, pp. 187-195, 1467-9876

Kolenikov, S. and Angeles, G. (2009). Socioeconomic Status Measurement with Discrete Proxy Variables: Is Principal Component Analysis a Reliable Answer? *Review of Income and Wealth*, Vol. 55, No. 1, 128-168.

Kula, J. (2004). AIC and BIC: Comparisons of Assumptions and Performance, *Sociological Methods & Research*, Vol.33, No.2, pp. 188-229, 1552-8294.

Laaksonen, M., Rahkonen, O., Karvonen, S. & Lahelma, E. (2005). Socioeconomic Status and Smoking: Analysing Inequalities with Multiple Indicators, *European Journal of Public Health*, Vol.15, No.3, pp. 262-269, 1464-360X.

National Cancer Institute . Surveillance Epidemiology and End Results. Standard populations - 19 age groups. http://seer.cancer.gov/stdpopulations/stdpop.19ages.html (accessed 09 Mar 2011

Philips BU, Jr., Gong G, Hargrave KA, et al. Correlation of the Ratio of Late-stage to Non-Late-stage Cancer Cases with the Degree of Socioeconomic Deprivation among Texas Counties. International Journal of Health Geographics 2011;10: 12.

Pomeroy, R.S., Pollnac, R.B., Katon, B.M., & Predo, C.D. (1997). Evaluating Factors Contributing to the Success of Community-Based Coastal Resource Management: The Central Visayas Regional Project-1, Philippines. *Ocean and Coastal Management*, Vol. 36, No. 1-3, 97-120.

Richardson LC, Royalty J, Howe W, et al. Timeliness of breast cancer diagnosis and initiation of treatment in the National Breast and Cervical Cancer Early Detection Program, 1996-2005. Am J Public Health 2010, 100:1769-1776.

Ruel, M.T. and Menon, P. (2002). Child Feeding Practices Are Associated with Child Nutritional Status in Latin America: Innovative Uses of the Demographic and Health Surveys. *Journal of Nutrition*, Vol. 132, No. 6, 1180-1187, 1541-6100.

Salmond C, Crampton P, Sutton F. NZDep91: A New Zealand index of deprivation. Aust N Z J Public Health. 1998;22:835-837

Schwartz, G. (1978). Estimating the Dimension of a Model, *Annals of Statistics*, Vol.6, No.2, pp. 461-464, 0090-5364

Singh GK. Area deprivation and widening inequalities in US mortality, 1969-1998. Am J Public Health. 2003;93:1137-1143.

Singh GK, Siahpush M. Increasing inequalities in all-cause and cardiovascular mortality among US adults aged 25-64 years by area socioeconomic status, 1969-1998. Int J Epidemiol. 2002;31:600-613.

Texas State Data Center. Published Reports. http://txsdc.utsa.edu/Reports. accessed 13 Oct. 2011.

Texas Department of State Health Services . County supply and distribution Tables. http://www.dshs.state.tx.us/chs/hprc/PHYS-lnk.shtm (accessed 17 Mar 2011).

Tobin, J. (1958). Estimation of Relastionships for Limited Dependent Variables, *Econometrica*, Vol.26, No.1, pp. 24-36, 1468-0262

U.S. Census Bureau. 2005-2009 American Community Survey 5-Year Estimates. http://factfinder.census.gov/servlet/DCGeoSelectServlet?ds_name=ACS_2009_5Y R_G00_

Vyas, S. and Kumaranayake, L. (2006). Constructing Socio-Economic Status Indices: How to Use Principal Component Analysis. *Health Policy and Planning*, Vol. 21, No. 6, 459-468, 1460-2237.

Ward E, Jemal A, Cokkinides V, Singh GK, Cardinez C, Ghafoor A, Thun M. Cancer disparities by race/ethnicity and socioeconomic status. CA Cancer J Clin. 2004 Mar-Apr;54(2):78-93

World Health Organization. Commission on Social Determinants of Health. Closing the gap in a generation: Health equity through action on the social determinants of health. http://www.who.int/social_determinants/thecommission/finalreport/en/index. html (accessed 25 August 2011)

Public Parks Aesthetic Value Index

M. K. Mohamad Roslan and M. I. Nurashikin
Universiti Putra Malaysia
Malaysia

1. Introduction

1.1 Aesthetic

Aesthetics is a branch of philosophy dealing with the nature of beauty, art, and taste with the creation and appreciation of beauty. It is more scientifically defined as the study of sensory or sensory-emotional values. Broadly, scholars in the field define aesthetics as critical reflection on art, culture and nature.

Traditionally, aesthetics is specific to the philosophy of art on level of beauty of any object but later, it was expanded to include the sublime with careful analyses on the characteristics. Studies on the aesthetics level experience and its attitude being conducted had lead to a survey of the scope of aesthetics together with an account of redefinition of the term.

According to Kant (1964), the theory of beauty has four clear aspects which are freedom from concepts, objectivity in the judgment, the disinterest of the viewer, and evenness in obligation. Freedom from concepts refers the cognitive powers of human understanding and imagination judge in identifying an object, such as stone. In this situation, the replacement of a certain individual object by a group of object would bring the experience of pure beauty when the cognitive powers are held to be in harmonious. Consequently, this results the second aspect that is the objectivity in the judgment since the cognitive powers are common to all who can classify that group of object. However, this was not the basis on which the apprehension of pure beauty was obligatory. According to Kant (1964), it derived from its disinterest because pure beauty not necessarily brings the fully agreement of the visitor to its aesthetic level nor does it raise any desire to possess the object but it just make satisfaction in a distinctive intellectual way.

1.1.1 Aesthetic value

Aesthetics value can be defined as theory of the level of beauty of a certain natural resources (Slater, 2006). This term appear when there was some interest among researchers to assess the level of aesthetic quantitatively instead of just qualitatively. It is about the objectivity and universality of judgments of pure beauty. It can also be defined as the individual judgment of the quality of beauty.

The main debate over aesthetic value concerns on social and political matters, in many different points of view. The central question concerns whether there are existing aesthetics expertises that have aesthetic interests and whether their view represents a fair view since, from a sociological perspective, it only just a portion of the whole population.

1.1.2 Environmental aesthetics

This paper would explain on public parks which are related to environment. In this case, according to Fisher (2003), environmental aesthetics applies naturally in the study of the aesthetics of nature conducted under the influence of environmental aspects concerns. The rapid growth of concern for the natural environment over the last third of the 20th century has brought the welcome reintroduction of nature as a significant topic in aesthetics such as the aesthetic interactions with nature, the aesthetic value of nature, and the status of art about nature. Nature began to be seen as comprising landscapes compelling in their own wild beauty and objects valuable in their smallest natural detail.

Environmental aesthetics consists of two terms which are environmental roots that emphasize on the aesthetics of nature, and aesthetic appreciation that focusing on the notion of environments of all sorts as objects of appreciation (Fisher, 2003). Based on these, nature is regarded not as an object to be exploited, but as something with an autonomous and worthy existence in itself. A thing is right when it tends to preserve the integrity, stability, and beauty of the biotic community but is wrong if otherwise.

According to Berleant (1998), environmental aesthetics applies to the researches on the aesthetics of nature of all sorts of environments either human made or natural conducted under the influence of environmental concerns. This environmental aesthetics also incorporates city planning, landscape architecture and environmental design, and it is significant because, whether applied to nature or built environments because it directly associating the modern structure or design to the standard theories in aesthetics. In short, Thompson (1995) found that although the authorities have an obligation to preserve beautiful high aesthetic value parks, they also have an obligation to preserve aesthetically valuable nature areas.

1.1.3 Aesthetic value judgment

Judgments of aesthetic value rely on our ability to discriminate at a sensory level. Aesthetics examines our affective domain response to an object or phenomenon. According to Slote (1971), a personal view on the level of aesthetic cannot represent the view of the community because each individual in the community also have their own view unless all of them have the same perception.

Since beauty is not a tangible object, judgment of aesthetic value is subjective to any individual perception. For example, it can be related to desirability or expectation on certain aspects of the nature. This assessment would also be based on many other kinds of issues such as senses, emotions, intellectual opinions, culture, preferences, values and behavior (Slater, 2006).

Aesthetic judgment is not only focusing on sensory discrimination but also linked to capacity for pleasure (Kant, 1964). The sensory discrimination that raising the pleasure would results the enjoyment. However, in order to judge something as having high aesthetic value, it needs a third requirement, that is it must involve the viewer's capacities of reflective contemplation. In short, judgments of aesthetic involved sensory, emotional and intellectual all at once.

1.1.4 Factors of aesthetic value judgment

There are some factors that could contribute in viewer view or perception on aesthetic value level. Negative reaction such as disgust shows that sensory detection is linked to body language, facial expressions, and behaviors (Slote, 1971). Aesthetic judgments also linked to emotions that partially embodied in physical reactions. For example, watching a well-design landscape parks may give the viewer a reaction of awe and consequently, physical reactions.

In the other hand, aesthetic judgments may be affected by culture or norm of the community. Other factors that could affect the judgments are desirability, economic, political, or moral value.

Aesthetic judgments can be highly parallel or highly contradict between viewers (Mothersill, 2004). In the other hand, aesthetic judgments were partly intellectual and interpretative. Therefore aesthetic judgments were usually based on the senses, emotions, intellectual opinions, will, desires, culture, preferences, values, subconscious behavior, conscious decision, training, instinct, sociological institutions, or combinations.

1.2 Index

Index is a numerical standardized value of evaluation of a variable that is in composite form, meaning that this variable was formed through the integration or combination of several other variables.. The evaluation process of this composite variable is not an easy process since there is no standard value used as a base of comparison of the evaluation. Therefore the indices are the best way to be introduced to determine that particular standard value.

A Public Parks Aesthetic Value Index, proposed in this paper is a measurement that aggregates numerous aesthetic value indicators into one consolidates and objective value representing the status of aesthetic value in public parks in relation to some specific use. The mode of assessment was carried out with a full consideration on certain criteria as suggested by Haslina (2006).

1.3 Public parks

Public parks is a bounded area of land, either in the form of natural or man-made state, dedicated to recreation use and generally characterized by its natural historic or landscape features. It is used for both passive and active forms of recreation and may be designated to serve the residents of a neighbourhood, community, region or province. This area is maintained by a local or private authority for recreation which can stimulate the morale and transform the cities to become more attractive and entertaining to visitors (Cranz, 1982).

Parks commonly resembles a certain formation of landscaping which any individuals who visit it would find more relaxing and feeling enjoyment of any recreation activities, for example picnics and sporting activities. Many smaller neighbourhood parks are receiving increased attention and valuation as significant community assets and places in heavily populated urban areas.

Traditionally, public parks are classified by their horticulture design, recreational value and public friendliness. It often provide leisure and sports areas for children and teenagers. It may also equip with semi natural or man-made habitats such as woodlands, heaths and wetlands. The formal garden designs and high composition of flora species characterized the horticultural history of public parks.

Rapid population growth rate and prolonged economic expansion consequently results a high degree of urbanization which could led to serious degradation of living environment and public amenities including air and water pollution, expanding temperature diffusion and decrement of public open space areas.

In order to have the functioning public parks in giving relaxation, enjoyment and pleasuring to the community, besides a good management and proper location, public parks should have a good aesthetic value. Aesthetic value for public parks can be defined as the level of beauties of natural beauty of public parks. However, the assessment of aesthetic value cannot be done in a straight forward approach because as mention earlier, this term is subjective to qualitative variables which are in composite form that means it consists of several aspects that need to combine together. Therefore the index of aesthetics value needs to be applied and measure to assess the aesthetic value of the parks. Index is a numerical standardized value of the composite form.

1.3.1 Characteristics of aesthetic values in public parks

Aesthetic values in Public Parks are normally characterized with five scopes; tree, fauna, lake, flower and building (Haslina, 2006).

1.3.1.1 Tree

The basic value of the tree in public parks could be based on emotion, aesthetic or strictly utilitarian. However, visitors seldom perceive the value as strictly aesthetic or monetary gain. In some areas, public used to spend on tree care and management which reflects the approximate value of trees. Trees may be selected to be planted in the public parks based on texture, fragrance, size, shape and colour. The replanting of best and cost effective tree species could create beauty environment and pull the visitors to come.

1.3.1.2 Fauna

Fauna is one of the aesthetic aspects in public parks. The existing of fauna species such as birds and fish would make the parks more harmonious, relaxing and enjoyable. Fauna species could revitalized flora survival needs and hence support each other. In the other hand, the abundance of greens in public parks could cause a significant increment in wildlife population.

1.3.1.3 Lake

Public parks, especially in urban areas are normally provides the visitors with lake as a choice for recreation. The physical attractiveness of a particular area contributes to its aesthetic value. The well-managed lake would provide the visitors not only the natural visual element but also a serene sound and smell. These aesthetic characteristics of the lake make up the reason that visitors might find lake as beautiful and attractive.

1.3.1.4 Flower

The function of flowers in public parks is well-documented. The evolving, flourishing process of civilization has increased the usage of flower in the form of civility, art and religion. For example, the Japanese people regarded flowers as materials for spiritual expressions (Haslina, 2006).

1.3.1.5 Building

Besides location, the appropriate design for the intended function of the building is one of the important elements of value The style and functional utility of a building are necessarily interrelated to create a success. In the valuation process, the aspects that are critical are functionality of the layout, design attractiveness, the appropriateness of the material and its quality and workmanship applied.

1.4 Elements of design aesthetic value

According to Janick (2010), there are four main design elements which become the visible features of all objects. They are as follows.

a.   Colour – Refer to visual sensation produced by different light wavelength that may be described in term of its hue, intensity and chroma.
b.   Texture – Refer to the visual effect of tactile surface qualities.
c.   Form – Refer to the shape and structure of a three dimensional object
d.   Line – Refer to one dimensional interpretation of form that delimits the shape and structure

1.5 Aim of study

Non-existence of this public parks aesthetic value index in Malaysia brings the issue of the uncertainty whether the existing public parks are functioning well in giving relaxation, enjoyment and pleasuring to the community and visitors. This study aims to introduce an index to assess the level of aesthetics value of public parks and applied it to assess the aesthetic value level of a certain public parks in Malaysia.

2. Methodology

2.1 Study area

This research had been conducted in Shah Alam Lake Garden, Selangor, Malaysia which is one of the popular recreational public parks among domestic tourists in Selangor (Fig. 1). This garden, consists of three man-made lakes is located in the city centre and provides the community and visitors with water-sports besides some basic leisure activities. This urban public parks, managed by government has a beautiful, green and unique landscape with the presence of many different bird species.

2.2 Data collection

Due to the assessment of aesthetics value is subjective to individual's perception as mention earlier, the first stage of this research is a qualitative study with questionnaire as a tool of data collection.

Fig. 1. The map of Shah Alam Lake Garden

The field survey was conducted on weekdays, weekend and public holidays. The respondents involved those who were present at the lakes at the time when questionnaires were given. Besides that, the respondent age of above 16 years old was also been point out in order to ensure that the respondents could give the grading to the aspects fairly. The questionnaires were designed based on the information obtain through discussions with experts. The questions provided in the questionnaires were straight forward and highly focusing on the aim that was to assess the respondent perception on the aesthetic level of the parks.

This questionnaire has two parts, Part 1 and Part 2. Part 1 involved the ranking of importance of five aspects in study while Part 2 involved the ranking of aesthetic value of each aspects attribute. Likert scale was used as aesthetic value grading scale for each sub-aspect from 1 to 10 where 10 means very high value. Moreover, Likert scale was also used as importance scale for each aspect from 1 to 5 where 5 means not important at all. Five aspects in study were as follows.

a.   Aesthetic value of Tree

After the final round of discussion with experts, this section focused on the form, colour, shape, scent, sound, appearance, function, site selection, and quantities and diversity of species that could influencing the visitors to visit the park.

b.   Aesthetic value of Animal or Fauna

After the final round of discussion with experts, this section focused on the quantities and diversity of species, sound, disturbance, smell, colour, shape, habitat and its function in the parks.

c.   Aesthetic value of Lake

After the final round of discussion with experts, this section focused on water quality in term of presence of fish, colour, lake design, sound, disturbance, odor, site selection and its function to visitors.

d.   Aesthetic value of Flower

After the final round of discussion with experts, this section focused on type of species, shape, form, colour, sound, aroma, site selection, design, quantities and its function.

e.   Aesthetic value of Buildings

After the final round of discussion with experts, this section focused on design, level of cleanliness, material quality, number of buildings, suitability of the location and its functionality.

2.3 Sampling technique

Sampling was conducted using convenience sampling techniques. In convenience sampling, the sample selection is based on availability and accessibility. This technique is quite suitable in this study due to the high mobility of visitors and also unknown activities among visitors.

After calculations of sample size based on the population size obtained, it was found that as many as 407 respondents need to be taken as a sample size from the whole population. Although convenience sampling technique was applied, the sample of respondents was still taken from the whole area of the lake to ensure that the results would represent the whole visitors of the lake from all areas. Although the likelihood of the sample being unrepresentative of the visitor population would be quite high, the information could still provide a fairly significant insight and become a good source of data in exploratory research.

2.4 Data analysis

Second stage of this research involved quantitative study where all the qualitative data being analyzed and quantify to give a certain value to indicate the aesthetic level. Descriptive analysis was first used to determine the importance of each aspect in determining the aesthetic level. Table 1 shows the results of the analysis. It was found that the most important aspect for the respondents to visit the parks was tree with mean value of 2.00 followed by lake, flower and fauna with mean value of 2.34, 3.06 and 3.33 respectively. The aspect of buildings was considered as not important to be including in the assessment with the mean of 4.25. From the observations and feedbacks from visitors, the buildings in the parks doesn't have any uniqueness that can be consider as important enough to attract them to visit. Therefore, the aspect of building would not be analyzed in the second stage.

Aspects	Mean	Median
Tree	2.00	2.00
Fauna	3.33	3.00
Lake	2.34	2.00
Buildings	4.25	5.00
Flowers	3.06	3.00

Table 1. Descriptive analysis of the aspects in study

2.4.1 Principal Component Analysis (PCA)

PCA is a technique for simplifying a data set by reducing multidimensional data sets to lower dimensions for analysis. It is an orthogonal linear transformation that transforms the variance covariance matrices to a new coordinate system such that the greatest variance would lie on the first principal component and so on while at the same time still retaining the characteristics of the data set.

The correlation between variables or aspects and component, indicated by factor loading was used as a basic for classifying the dominant variables in each component. If the factor loading value is more than 0.7, the attribute can be considered as dominant and plays a main role in that component (Nunnally, 1978; Hair, 2009) because it would account for more than 50% of the variance in all variables.

The next step of analysis was the usage of Principal Component Analysis (PCA). The reason of applying this analysis was to extract the main attributes from all attributes of each aspect. This is important as in index development, the number of attributes must be optimum in order to represent a good indicator for assessment tools. The index with a lot of attributes would seldom represent a good status indicator. Moreover, another important step in developing or creating an index is the selection of only the most important or dominant attributes as these attributes play a big role in determining the status of the variable such as the status of aesthetic value of the public parks in this study. Since in PCA, the first component is always dominant compare to the other component, the selection of the main attributes was done only in first component.

However, this dominant component cannot be chose straight away as the indicator used in index as there is still some effect of correlation between components. According to Hair (2009), factor loading is the coefficient used to interpret the role of each variable and factor and the unrotated solution may not provide a true pattern of variable loadings due to the presence of this effect. In order to solve this problem, the whole extracted component need to be rotated. According to Rencher (2002), the most popular rotation technique is the varimax technique which seeks rotated loadings that maximize the variance of the squared loadings in each component. Indirectly, this varimax rotation could eliminate the effect of correlation between components.

As mention earlier, only four aspects were being considered after the aspect of building had been omitted. In the preliminary round, PCA was run separately for all five aspects to determine the main or dominant attributes in each aspect. As mention earlier, the index is the composite variable that was formed through the combination of several other variables. Mohamad Roslan et. al (2007) found that the index that can best explaining the real condition or situation is the index that focused on dominant variables in each scope or classification. In this study, it refers to dominant variables in the scope of tree, fauna, lake and flower.

Table 2 shows the component structure of trees after the varimax rotation process of PCA. The Table reports the factor loadings for each variable on the components between the item and the rotated factor. In this case, the factor loading value of 0.7 was chosen as as threshold value explained earlier. It was found that the attributes of maintenance, species selection and form has the factor loading value of 0.773, 0.766 and 0.741 respectively which are higher than the threshold value.

Attributes	Component	
	Component 1	Component 2
Form	0.741	-0.027
Colour	0.425	0.419
Maintenance	0.773	0.101
Quality	0.202	0.699
Species selection	0.766	0.031
Function	0.058	0.840
Quantity	0.499	0.376
Location	-0.078	0.762

Table 2. Results of PCA analysis for the aspect of tree

From the feedback of the visitors, different form of trees can create an attractive view to the park while giving an aesthetic visual to them. At the same time, a proper tree maintenance existed in the park becomes an added value in term of aesthetic value.

Table 3 shows the results of PCA analysis for the aspect of animal or fauna. It was found that higher factor loading recorded for habitat and selection of species with 0.872 and 0.839 respectively. From the feedback of visitors, the varieties of species play a pull factor to attract them to the parks.

Attributes	Component	
	Component 1	Component 2
Species selection	0.839	-0.025
Quantity	0.503	0.490
Habitat	0.872	0.094
Sound	0.067	0.779
Cleanliness	0.663	0.260
Function	0.095	0.843

Table 3. Results of PCA analysis for the aspect of animal or fauna

Table 4 shows the results of PCA analysis for the aspect of lake. There were two attributes that have the value of factor loading more than threshold, colour and quality with 0.905 and 0.903 respectively. From the feedback of the visitors, both attributes are important in influencing their first impression.

Attributes	Component	
	Component 1	Component 2
Quality	0.903	0.136
Design	0.309	0.712
Colour	0.905	0.162
Cleanliness	0.575	0.524
Sound	0.261	0.650
Function	-0.041	0.851

Table 4. Results of PCA analysis for the aspect of lake

Table 5 shows the results of PCA analysis for the aspect of flower. It was found that only one component could be form from the analysis and from this component, five attributes were major attributes. They were quality, odour, quantity, colour and species selection. These five attributes were related to each other because if choices of species were good enough, it reflected the qualities, quantities, colour and odour. In short, a good quality species produced a good flower.

Attributes	Component
Species selection	0.746
Quantity	0.773
Quality	0.810
Colour	0.754
Odour	0.796
Function	0.586

Table 5. Results of PCA analysis for the aspect of flower

From all these Table 2 to Table 5, it was found that most major attributes that could be a major pull factors for the visitors to visit were in the aspect of flower. In the final analysis of PCA, all main attributes in each aspect were analyzed together. In this stage, the data was separate into two parts, 300 of them in one part, used to determine the main attribute in exploratory while another 107 were used in validating the exploratory results. This process was done because the results of this stage would be use in creating the index and therefore need to be validating to ensure that the index is valid to be use. The data set of 300 respondents still covering the whole public parks areas and the same applied for the data set of 107 respondents.

Table 6 shows the results of PCA analysis for data set in determining the main attributes in exploratory. It was found that odour of flower, colour of water, quality of flower, quantity of flower, water quality and colour of flowers were the main attributes from all selection attributes, having the factor loading value of 0.805, 0.766, 0.756, 0.734, 0.704 and 0.700 respectively.

Attributes	Component	
	Component 1	Component 2
Tree formation	0.141	0.800
Tree maintenance	0.273	0.771
Tree species selection	0.175	0.727
Water quality	0.704	0.255
Colour of water	0.766	0.177
Flower species selection	0.679	0.164
Quantity of flower	0.734	0.186
Quality of flower	0.756	0.175
Colour of flower	0.700	0.172
Odour of flower	0.805	0.172

Table 6. Results of PCA analysis for data set to determine the main attrbutes

Table 6 shows that from the exploratory results, it was found that six attributes could be taken as aesthetic value index indicators from 26 attributes. They were odour of flower, colour of water, quality of flowers, quantity of flowers, water quality and colour of flowers following the order. This order or arrangement is important in the next stage of analysis. Consequently, running the same process for the validation data set gave the same arrangement of attributes though the factor loading value is not same.

2.4.2 Benchmarking analysis

Benchmarking analysis is the analysis used in comparing the current project, methods or processes with the best practices in order to drive improvement. The objective of this analysis is to set an appropriate reliability and quality matrices for the output based on similar present products.

This method had been used by Mohamad Roslan et. al (2007) in their study on the creation of ground water quality index. They found that this method is suitable and highly sensitive. However, in their study, they used the quantitative data of several prospect water quality indicators.

After the analysis of PCA, the next stage is to develop the index. This study would offer some uniqueness in the index formation; the index is not in statistical equation but in graphical feature (Figure 3). Each vertex represents the attributes, extracted from the PCA and arranged in clock-wise order. This graphical feature which is part of benchmarking analysis was drawn to calculate the index with the help of mean analysis. From this graphics, the index was calculated as the percentage of the polygon area drawn using the data to the whole polygon area as shown in the green area in Figure 3.

Procedure of developing the index using Benchmarking Analysis is as follow.

1. Fix the value of analysis with the scale of 0 to 10 for each main attribute extracted from PCA
2. Built the polygon
3. Calculate each triangle (Fig 2) area inside the polygon drawn based on the value from data using formula (1).

$$A=0.5 \times a \times b \times \sin \theta \tag{1}$$

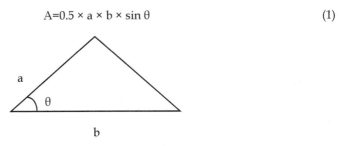

Fig. 2. Triangle in the polygon

4. Determine the ratio of the area using formula (2).

$$IV=A/L \times 100 \tag{2}$$

WhereIV=Index value

A= Total triangle area

L=Total polygon area

5.   Evaluate the aesthetic value using the index.

2.4.3 Sensitivity analysis

This analysis was also conducted to the model to determine its sensitivity to the change of the value. For example, determine the changing rate of the index when any attribute fall from 10 to 9 in benchmarking scale. In this case, it was found that the index value would fall as much as 3.33% for any single attribute that fall as much as one unit in benchmarking scale.

2.5 Creation and calculation of the index in the study

Before the polygon can be drawn, the mean analysis for the main attributes extracted from PCA need to be conduct to determine the level of aesthetic level for each attribute within the Benchmarking scale of 0-10. This analysis involved all respondents without separation. Table 7 shows the mean analysis for six main attributes. It shows that during the study, the attribute of quality of flower has the highest ranking compare to the other five attributes with the value of 5.12.

Attributes	Mean
Water quality	4.13
Colour of water	3.99
Quantity of flower	4.75
Quality of flower	5.12
Colour of flower	4.90
Odour of flower	4.28

Table 7. Mean analysis for six main attributes

The values in Table 7 were used in drawing the polygon and consequently the Public Parks Aesthetic Value Index. Fig. 3 shows the polygon used to calculate the index with the scale of 0-10. Using the procedure and formula as mentioned earlier, it was found that the value for Shah Alam Lake Garden Aesthetic Value Index is 20.44.

This study also furnished the index with the grading system; develop using the concept of the polygon area to help in grading the aesthetic value index through benchmarking analysis. Five grades were applied in this product that is excellence, good, satisfactory, poor and very poor (Table 8). These grades were set up using the benchmarking level scale of 9, 7, 5. 3 and 1 for all six attributes.

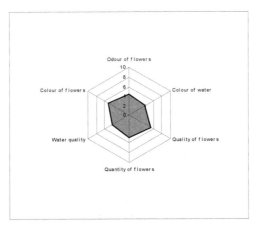

Fig. 3. The index with the mean value for each variables

Grades	Value range
Excellence	81.00 – 100.00
Good	49.00 – 80.99
Satisfactory	25.00 – 48.99
Poor	9.00 – 24.99
Very poor	0.00 – 8.99

Table 8. Grading system

The final stage of this study is to apply the index to calculate the aesthetic value index of the study area. Based on the grading range (Table 8), the index value for Shah Alam Lake Garden Aesthetic Value Index of 20.44 can be classified as poor. From Figure 3 and Table 7, it was shown that almost all attributes were given grading scales less than 5 except the quality of flower. This contributes the overall grade level.

3. Conclusion

Taking into considerations the five criteria related to the assessment of aesthetic value index as mentioned by Haslina (2006) which primarily focuses on the beauties and the 'in' thing, this index would be a valuable tool in the management of aesthetic value in tourist attraction sites such as public parks, lake gardens and recreational forests. The minimal constraint in application made it suitably applied in assessing the aesthetic value under the sun.

This index was created with the data gain during daylight. Therefore, the main indicator variables in the index became dominant during daylight. However, the assessment for the night session may refer to another set of indicator variables. This brings the potential of future study on this aspect.

Another potential scope of future study is to use the quantitative method in measuring the value or grading for each attribute in the index instead of just collecting the data through a large size of respondent perception's. This could be done with the utilization of several high technology equipments.

4. References

Brady, E. (2005). *Aesthetics in Practice: Valuing the Natural World*. New York: City University of New York

Berleant, A. (1998). Environmental Aesthetics. In M. Kelly (Ed). *Encyclopedia of Aesthetics*. New York: Oxford University Press, 114-120.

Camp, J.V. (2005). *Freedom of Expression*. New York: The National Endowment for the Arts.

Carlson, A. (2000). *Aesthetics and the Environment: The Appreciation of Nature, Art and Architecture*. New York: Routledge.

Carlson, A. (2010). Environmental aesthetics. In E. Craig (Ed.), *Routledge Encyclopedia of Philosophy*. London: Routledge.

Cranz, G. (1982). *The Politics of Park Design: A History of Urban Parks*. Cambridge Press: Cambridge

Fisher, J. A. (2003). Environmental Aesthetics. In: the *Oxford Handbook of Aesthetics* ed. Levinson, J. Oxford: Oxford University Press

Hair, J. F., Black, W. C., Banin, B. J. & Anderson, R.E. (2009). *Multivariate Data Analysis*. 7th Edition. Prentice Hall: New Jersey.

Haslina, H. (2006). *Visitors Perception on Aesthetic Function of Ornamental Plant in Taman Tasik Titiwangsa, Kuala Lumpur*. UPM Faculty of Forestry Bachelor Thesis.

Hettinger, N. (2005). Allen Carlson's Environmental Aesthetics and the Protection of the Environment. *Environmental Ethics* 27(1): 57-76.

Janick, J. (2010). Plant iconography and art. *Source of information on horticultural technology, 1*.

Kant, I. (1964), *The Critique of Judgement*, trans. J.C.Meredith, Oxford University Press, Oxford.

Kent, A. J. (2005). Aesthetics: A Lost Cause in Cartographic Theory. *The Cartographic Journal*. 42(2). Pg 182-188

Lifestyle Information Network. (2007). Guidelines for Developing Public Recreation Facility Standards. Date of access: 20th October 2011.. http://lin.ca/resource-details/1477.

Mothersill, M. (2004). Beauty and the Critic's Judgment: Remapping Aesthetics. In *The Blackwell Guide to Aesthetics*. New Jersey: Blackwell Publishing Ltd

Mohamad Roslan, M.K., Mohd Kamil, Y., Wan Nor Azmin, S. & Mat Yusoff, A. (2007). Creation of a Ground Water Quality Index for an Open Municipal Landfill Area. *Malaysian Journal of Mathematical Sciences*. 1(2): 181-192

Nunnally, J. C. (1978). *Psychometric theory*. 2nd Edition. New York: McGraw-Hill.

Rencher, A.C. (2002). *Methods of Multivariate Analysis*. 2nd Edition. New York: John Wiley & Sons.

Slater, B.H. (2005). Aesthetics. Internet Encyclopedia of Philosophy. Date of access: 11 June 2011. http://www.iep.utm.edu/aesthetic.

Slote, M. A. (1971). The Rationality of Aesthetic Value Judgments. *The Journal of Philosophy*. 68 (22): 821-839

Thompson, J. (1995). "Aesthetics and the value of nature," *Environmental Ethics* 17, 291-305

Permissions

The contributors of this book come from diverse backgrounds, making this book a truly international effort. This book will bring forth new frontiers with its revolutionizing research information and detailed analysis of the nascent developments around the world.

We would like to thank Parinya Sanguansat, for lending his expertise to make the book truly unique. He has played a crucial role in the development of this book. Without his invaluable contribution this book wouldn't have been possible. He has made vital efforts to compile up to date information on the varied aspects of this subject to make this book a valuable addition to the collection of many professionals and students.

This book was conceptualized with the vision of imparting up-to-date information and advanced data in this field. To ensure the same, a matchless editorial board was set up. Every individual on the board went through rigorous rounds of assessment to prove their worth. After which they invested a large part of their time researching and compiling the most relevant data for our readers. Conferences and sessions were held from time to time between the editorial board and the contributing authors to present the data in the most comprehensible form. The editorial team has worked tirelessly to provide valuable and valid information to help people across the globe.

Every chapter published in this book has been scrutinized by our experts. Their significance has been extensively debated. The topics covered herein carry significant findings which will fuel the growth of the discipline. They may even be implemented as practical applications or may be referred to as a beginning point for another development. Chapters in this book were first published by InTech; hereby published with permission under the Creative Commons Attribution License or equivalent.

The editorial board has been involved in producing this book since its inception. They have spent rigorous hours researching and exploring the diverse topics which have resulted in the successful publishing of this book. They have passed on their knowledge of decades through this book. To expedite this challenging task, the publisher supported the team at every step. A small team of assistant editors was also appointed to further simplify the editing procedure and attain best results for the readers.

Our editorial team has been hand-picked from every corner of the world. Their multi-ethnicity adds dynamic inputs to the discussions which result in innovative outcomes. These outcomes are then further discussed with the researchers and contributors who give their valuable feedback and opinion regarding the same. The feedback is then collaborated with the researches and they are edited in a comprehensive manner to aid the understanding of the subject.

Apart from the editorial board, the designing team has also invested a significant amount of their time in understanding the subject and creating the most relevant covers. They scrutinized every image to scout for the most suitable representation of the subject and create an appropriate cover for the book.

The publishing team has been involved in this book since its early stages. They were actively engaged in every process, be it collecting the data, connecting with the contributors or procuring relevant information. The team has been an ardent support to the editorial, designing and production team. Their endless efforts to recruit the best for this project, has resulted in the accomplishment of this book. They are a veteran in the field of academics and their pool of knowledge is as vast as their experience in printing. Their expertise and guidance has proved useful at every step. Their uncompromising quality standards have made this book an exceptional effort. Their encouragement from time to time has been an inspiration for everyone.

The publisher and the editorial board hope that this book will prove to be a valuable piece of knowledge for researchers, students, practitioners and scholars across the globe.

List of Contributors

Louis Noel Gastinel
University of Limoges, University Hospital of Limoges, Limoges, France

Ferran Reverter, Esteban Vegas and Josep M. Oller
Department of Statistics, University of Barcelona, Spain

Tsai-Ling Liao and Chih-Jen Huang
Providence University, Taichung, Taiwan

Chieh-Yuan Wu
Department of the Treasury Taichung Bank, Taiwan

Ingunn Tho
University of Tromsø, Norway, Denmark

Annette Bauer-Brandl
University of Southern Denmark, Denmark

Érica C. M. Nascimento and João B. L. Martins
Universidade de Brasília, LQC, Instituto de Química, Brazil

Halina Kucharczyk and Kinga Stanisławek
Department of Zoology, Maria Curie-Skłodowska University, Lublin, Poland

Marek Kucharczyk
Department of Nature Conservation, Maria Curie-Skłodowska University, Lublin, Poland

Peter Fedor
Department of Ecosozology, Faculty of Natural Sciences, Comenius University, Bratislava, Slovakia

Donát Magyar
National Institute of Environmental Health, Budapest, Hungary

Gyula Oros
Plant Protection Institute of the Hungariain Academy of Sciences, Budapest, Hungary

Franc Janžekovič and Tone Novak
University of Maribor, Faculty of Natural Sciences and Mathematics, Department of Biology, Maribor, Slovenia

Elisabeth Stühler and Dorit Merhof
University of Konstanz, Germany

Eric Belasco
Montana State University, Department of Agricultural Economics and Economics, USA

Billy U. Philips, Jr. and Gordon Gong
Texas Tech University Health Sciences Center, F. Marie Hall Institute of Rural, Community Health, USA

M. K. Mohamad Roslan and M. I. Nurashikin
Universiti Putra Malaysia, Malaysia

Printed in the USA
CPSIA information can be obtained
at www.ICGtesting.com
JSHW011405221024
72173JS00003B/427